Price Yourself Right

PRICE YOURSELF RIGHT

*A Guide to Charging
What You're Worth*

JANE FRANCIS

iUniverse, Inc.
New York Lincoln Shanghai

Price Yourself Right
A Guide to Charging What You're Worth

Copyright © 2006 by Jane Francis

All rights reserved. No part of this book may be used or reproduced by any means, graphic, electronic, or mechanical, including photocopying, recording, taping or by any information storage retrieval system without the written permission of the publisher except in the case of brief quotations embodied in critical articles and reviews.

iUniverse books may be ordered through booksellers or by contacting:

iUniverse
2021 Pine Lake Road, Suite 100
Lincoln, NE 68512
www.iuniverse.com
1-800-Authors (1-800-288-4677)

The moral right of the author has been asserted.
Condition of sale
This book is sold subject to the condition that it shall not, by way of trade or otherwise, be lent, resold, hired out or otherwise circulated without the publisher's prior consent in any form of binding or cover other than that in which it is published and without a similar condition including this condition being imposed on the subsequent purchaser.
Professional indemnity
This publication is sold with the understanding that the author and publisher are not engaged in rendering personalized professional services. If professional advice or other expert assistance is required, the services of a competent professional person should be sought.

ISBN-13: 978-0-595-38601-7 (pbk)
ISBN-13: 978-0-595-82982-8 (ebk)
ISBN-10: 0-595-38601-6 (pbk)
ISBN-10: 0-595-82982-1 (ebk)

Printed in the United States of America

To Linda, who asked me to write this book. Thank you for the conversation that went like this:

Me: "There's no point writing a book on charging what you're worth. You just decide what you're worth, then you ask for it, and you either get it or you don't."

Linda, sighing: "But everyone knows it's not that easy."

Contents

Preface xi

Chapter 1:   Know Yourself 1

If other people can charge what they are worth, why don't you? 2
What do you have to gain by charging less than you are worth? 2
What else is stopping you? 3
Were you brought up to expect less than you are worth? 3
Is it because you don't think you are anything special? 4
Are you too willing to please? 5
Do you allow yourself to be belittled or bullied? 5
Are you letting your feelings get in your way? 7
Is it because your thoughts are faulty? 9
Are you identifying too much with your client? 10
Are you not charging your worth because your client is a friend? 10
Does your client expect to get it for nothing? 12
Is your client a charity? 14
Do you lack confidence? 14
Are you always giving away discounts? 15
Do you not charge enough because you are not being businesslike? 15
Do you underestimate the amount of work involved? 16
Are you worried about the consequences? 17
Do you think you don't have enough time? 17

Chapter 2: Understand Your Relationship to Money 18
It takes one to know one: think about your own style 19
How closely do you identify with money? 21
Are you master of your money, or is it master of you? 22
Will you ever have enough? 22
Money and opinions—yours and mine 23
Are you conflicted about profit? 24
The illusion of rationality 25

Chapter 3: Know Your Client 28
What do you know about your client's money mentality? 29
What clients say and what they really mean 30
Educate your customers 31
Provide excellent value 32
Interpret your customer's desire for "more prettier" 32
What sort of clients do you want to attract? 33
Not all customers are created equal 34

Chapter 4: Know Your Options 36
Perception is reality 36
What can your customer afford? 37
Customer segmentation 38
Consider your costs 38
Price with an eye on your competition 38
Market rates 39
Supply and demand 39
Price and volume 40
Price to suit the changing times 40
Usage rights 41
Learn from the alpaca breeders 41
The value of scarcity 41
Put a price on time 42
Take a commission or a percentage 42
Put a price on your brilliance 42

Retainer and royalties *43*
Your client will be glad to pay for your experience *43*
Price to save your clients the bother of doing it themselves *44*
Price to meet objectives *44*
Price according to risk *44*
Price according to providence *45*
Seek the highest bidder *45*
Price to manage peaks and troughs *45*
Offer a free sample *46*
Suck it and see *47*

Chapter 5: Determine Your Benchmarks 49

About pricing time *51*
About setting a profit revenue goal *52*
Your benchmark figure *54*
Forecast your fixed costs *54*
How many hours do you plan to work? *55*
What's your break-even? *56*

Chapter 6: Look Out for Number One 58

Should you wholesale your time? *59*
What units of time should you charge? *60*
Should you charge by the job, hour, or percentage? *61*
How many customers can you manage? *63*
What will the market bear? *63*
The value chain *64*
Experiment with price points *65*
What attributes or assets can you leverage to increase perceived value? ... *66*
What can you leverage for profit? *67*
Work out the cost of your conversions *67*
Promotional pricing and repeat business *67*
The truth about discounting *68*
Ten easy ways to discount and lose money, FAST! *69*
Twenty tips for negotiating success *70*

Chapter 7: Decide Your Value 75
Ask your customers 76
Get a second opinion 76
Now get a third, fourth, and fifth opinion 77
Test the market 77
Decide what suits you 77

Chapter 8: Present Your Price with Confidence 79
It's not what you say, but how you say it 80
Practice, practice, practice 80
Have your answer ready 81
Be prepared for the difficult moments 83
But what if they object to your price? 84
Perhaps you're not compatible 86
Facing rejection 87
What to do if the client gets personal 87
How to respond if they ask for a discount 88
What if your client tries to undermine your competence? 89
How to deal with friends who want to work with you 90
How to respond when you are "put on the spot" 91
How to manage a client who quibbles over the bill 93
What to do when confronted with a "deal-breaker" 93

Chapter 9: Focus on Your Dream 95
Know that you are worth it 96
Give yourself encouragement 96
If someone else can do it better than you, let them! 97
Wealth is a journey, not a destination 97
The journey to a million starts with one 98

Recommended Reading 99
Glossary of Terms 101

Preface

If I was reading this book, wanting to learn how to price myself right and charge what I'm worth, I'd be pretty skeptical. "How can you do that?" I'd ask. "You don't even know me or my business." Which is perfectly true; I don't.

But one thing I do know is that whatever your position in life, you will end up dealing with people and asking them to pay you—and I can help you work toward being as professional (and relaxed) about asking for the money as you are at earning it.

The truth is that pricing yourself right and charging what you are worth is as much about psychology as it is about business tactics and money.

It is also about changing old habits and finding and practicing new ones that work for you.

If you're not already charging what you are worth, think about how much more you could earn and how much better you'll feel about yourself when you do. Visualize the positive flow-on effect your good example will set for friends and family, who will also enjoy your higher self-esteem.

For the past fifteen years, I have been a consultant; this means I have had to negotiate my "salary" every time I take on a new job.

I have learned from experience how to undercharge and how to charge what I'm really worth. To paraphrase Sophie Tucker, there have been times in my life when I've had more cash flow, and times when I've had less—and more is definitely better.

I variously describe my business as either "an exclusive direct-marketing advertising agency" or "freelancer," depending on my self-esteem on the day and who is asking. Regardless, I have created advertising and direct-marketing campaigns for many international pharmaceutical products. I have also worked on campaigns in the insurance and banking, property, and investment industries, and I was the source of direct-marketing creative work for an international airline.

Early on, in David and Goliath style, I established a reputation within the industry by winning five gold medals plus the "best of show" award for a direct-marketing campaign—and all this in a business I began just after my first child was born. I now have four children, and I have never made any pretence about working from a home office, despite pressure to conform to the corporate way of doing things.

It's not easy being in business for yourself; not only do you have to be good at making the product or service that you offer, but you also need to excel at business in order to *stay* in business.

I've never yet found a university course that deals with my type of real life, so the lessons here come from the University of Self-Employment—they are what some might call "street smarts," and while some of those streets turned out to be culs-de-sac, others turned into major highways.

In this book, I'll teach you what I've learned about price-setting. I'll also help you figure out how much you should charge and give you the skills and confidence to charge what you are worth—and feel good about it.

Acknowledgments

Special thanks to all the friends and clients who have allowed me to learn from them. Please note that in all case studies and examples throughout this book, the names and identities have been altered to protect the privacy of each individual.

Chapter 1

Know Yourself

I can always sense when someone isn't confident about asking for money. There's usually a telltale hesitation in the voice, and words that come out as an apology.

Many of us have been brought up not to talk about money. Money is still a taboo subject, and we don't like to talk about it, so we don't. But if you want to claim what you deserve, it's no good saying, "I hate asking for money." You will have to learn how to change because it won't happen on its own.

Think about the last time you knew you should be charging more, but instead, sold yourself for less. What were the consequences? Here are some possibilities:

- You tired yourself out.

You accepted less than you are worth, so you have to earn more tomorrow, never catching up on yourself.

- You belittled your value.

You gave your customer an extra dose of time. You gifted them with a portion of your life, belittling the value you place on your life. Your time is your life.

- You devalued your worth.

You gave away free information. You undermined the value of your intellectual property or special skills that you invested time and money acquiring and developing.

- You exposed your lack of wealth consciousness.

You demonstrated your ignorance of or disdain for the power of money to accumulate, and compound, over time. The money you didn't make this time can't be put to work for you until you've earned it.

- You made it harder for yourself next time.

You denied yourself the chance to acknowledge your self-worth and you taught others to undervalue you. You signaled to your customer your availability to be exploited. "Here I am, going cheap! Be my guest—help yourself to the profits."

- You reinforced your own doubts about your self-worth, digging yourself deeper into the rut.

If other people can charge what they are worth, why don't you?

Logically, you might think as I did: "You just decide what you're worth, then you ask for it, and you either get it or you don't." But as my friend Linda so accurately replied, everyone knows it's not that easy.

Dr. Phillip McGraw, author of many insightful books, including *Life Strategies,* says, "People do what works."

Undercharging is all about psychology, and somewhere along the way, you must be finding a reward for this behavior; the only trouble is that the gain isn't monetary, when sometimes you'd really rather it was. Nice feelings don't pay the bills.

What do you have to gain by charging less than you are worth?

I can think of many ways you might be rewarded. For example, not charging what you are worth might make you feel safe by reducing the rejection you think you'd get if you charged more.

Perhaps undercharging allows you to feel like the "underdog," and that makes you feel justified when you complain to your family how your clients don't appreciate you. In this way, it buys you sympathy or gives you a sense of belonging within the group you associate with.

Here's another not very pleasant but possible reason you may undercharge. It provides an "out clause"; *i.e.,* you don't want to finish the job properly, so you think that if you leave it as it is and charge a bit less, "they'll understand."

Or perhaps, if you're not 100 percent confident in the job you've done, you might think: "Well, if it all falls apart, at least they didn't spend too much on it." Or: "What did they expect? It was only a hundred dollars."

If you see yourself in any of these situations, own up and learn a better way of communicating with your customers. Talk to your customer, so you are both clear about the details of the transaction. Perhaps that should include the option of them paying you less to do less, or you improving the quality and lifting the price.

What else is stopping you?

What other resistance do you have to charging your worth?

Try this exercise: Think of the job you're about to do. What are you going to charge? Now double that figure. How does that make you feel?

If it doesn't make you feel happy, excited, proud, or positive in some way, why not? Look for the blocks. You're not going to be motivated to change if doing so is going to be the cause of pain.

Were you brought up to expect less than you are worth?

What were your family or cultural beliefs? Do you have the "luck of the Irish," or do you come from a long line of hardship and strugglers?

Were you always told that no one with feet as big as yours would ever amount to anything? Or that you "couldn't possibly" move overseas, start your own business, sell the farm, or leave home? Was your brother always "the successful one" and were you "the useless one"?

Don't let this sort of baggage hold you back. Stop believing other people's doctrines, and start believing this right now: You are worth it.

Is it because you don't think you are anything special?

Have you ever started writing out a price quote, wanting to charge $1,000, but then ended up asking $500 because you listened to the voice inside your head? One that said: "What makes you think you are worth that much? Who do you think you are? You're not really that good."

Yet the customer may have been happy to pay $1,000 (or more!) for the job…but you undermined yourself. What do you have to gain from all this self-criticism?

Realize that thinking like this is not helping you. Tell your inner critic to *be quiet*, and uplift yourself with these thoughts:

"I do a worthwhile job."

"I am proud of what I do."

"I am good at what I do and worth what I charge."

"No one else can do it the way I do."

"I believe in what I do."

"I believe in myself."

If these don't feel right just yet, raise your standards and work to make them feel true for you.

I am always inspired by these words of Marianne Williamson, and I hope they may give you the courage to "let your own light shine":

"Our deepest fear is not that we are inadequate. Our deepest fear is that we are powerful beyond measure. It is our light, not our darkness that most frightens us. We ask ourselves, 'Who am I to be brilliant, gorgeous, talented and fabulous?' Actually, who are you not to be? Your playing small does not serve the world. There's nothing enlightened about shrinking so that other people won't feel insecure around you. And as we let our own light shine, we unconsciously give other people permission to do the same."

Are you too willing to please?

Are you the sort of person who'll go the extra mile by staying late at the office? Do you frequently do more than you were paid to do, and have been told you are "too nice for your own good"? How much is this costing you?

How many times have you told someone, "It'll only take five minutes," when the job will actually take more like an hour? If you said this to eight people, there goes a full working day—and probably most of your spare time in the weekend, as you try to catch up on other things because of the pressure you put yourself under.

Being a people pleaser is a very costly habit—I know!

Here are some possible reasons why you may do this:

You want the other person to like you.

You don't want the other person to think you are slow or incompetent.

You want to impress the other person with your great talents.

You want to save them from feeling a nuisance (when that's exactly what they are).

You feel guilty, if it was implied you hadn't done it correctly the first time.

Deep down, your competence feels challenged and you don't feel good enough.

If this is your issue, you need to realize that this money and pricing subject has got nothing to do with being nice, or being horrible; it's about doing things for money. If it's friends you want, you shouldn't have to buy them with favors. With the right techniques, it is possible to be nice *and* charge what you are worth.

Do you allow yourself to be belittled or bullied?

It is true that you teach what you need to learn. Though I have always believed you should charge what you are worth, there have been times I have compromised my beliefs. Growing up as the peacemaker in our family, I would do anything to avoid conflict. Fortunately, after I left home, I married the gentlest man I have ever known, but that still didn't stop me finding people to bully me in the workplace. This caused a lot of pain and stress, until I became aware of the ways I fell into the trap of playing out old ways to respond, giving a pay-off to the other person, not me.

In my first job upon graduation, I worked for a publisher who paid less than was required by law. The workplace had a high turnover of staff, with only the desperate staying any length of time. I stayed nine months because—and here's the pay-off—it was a stepping stone, giving me work experience and a reference to impress the next employer.

My underpaying employer had leverage over me not to take the matter further; nonetheless, because of my principles (it wasn't a great deal of money we were talking about), I took my grievance to the employment court anyway, knowing there were others working there that couldn't afford to risk their jobs.

I still vividly remember my nerves the day I turned up at the courtroom with my case officer from the Labor Department. I was petrified of confronting my immaculately attired employer, who was waiting outside the courtroom doors with his lawyer—who immediately greeted us by making an offer to settle out of court.

Desperately wanting to avoid the whole scene, I readily accepted, whereupon my ex-boss shook my hand, making eye contact for the first time ever. Relieved it was over, I hurried out of there, glad I had at least made him acknowledge his wrongdoing. I hardly taught him a lesson, but the lump sum was enough to pay for a windsurfer, a real luxury item to someone used to a student budget.

Looking back on it now, a part of me wishes that the just-out-of-university Jane had had the confidence to go through the courtroom doors and face the corporate suit. But another part of me sees the windsurfer as a lovely metaphor for that time in my life, an expression of the spiritedness of youth and freedom!

That man is now a multimillionaire many times over. It just goes to show that one woman's bully is another man's success story…which leads to my next point.

Bullies tend to build empires, so at some stage in your life, you will probably end up working for or doing business with one. You're therefore going to need to learn some effective techniques to deal with them or avoid them altogether.

Professional bullies often dress nicely, appear knowledgeable and successful, and hold positions of status. They will expect favors and demand that work be done unreasonably quicker, better, cheaper: "I don't care; just do it." Bullies often demand exclusivity or isolate you and limit your access to the "network," though they rarely reciprocate your loyalty.

Bullies are effective people-manipulators, brilliant at "reading" people and "pushing their buttons." They use techniques like leaving things to the last minute to block escape routes, and dangle carrots that won't be delivered, and often use "weasel words" (words like "virtually," "almost," and "possibly") as out clauses. They rely on a network of higher authorities—lawyers, accountants, politicians, and the like—but will avoid putting anything in writing.

Bullies look for clues when recruiting their "team." For example: At the end of a job well done, have you ever belittled your efforts by saying, "Oh, it was nothing, really," or "Anyone could have done it"? As insignificant as those statements may seem, trust me: a bully will notice!

Here's another signal: You've just told your client (or boss) that the job they want done will take you two days. She responds by looking surprised and saying, "But it only took Sue one day." You take two days, but only charge for one—bullies like it when you do that.

Be warned: with skills like that, you are bully bait! Interestingly, I've been told that the most effective way to address bullying in schools is to empower the victim. While it's not the victim's fault (that's got very little to do with it), in the long run the best strategy is to teach the victim how to disempower the bully. In just the same way, you need to learn how to restore the balance of power—or develop your bully "antennae" and leave these customers for someone else to deal with.

Are you letting your feelings get in your way?

Do you have some negative feelings that hold you back?

Imagine being led into a room filled with one-hundred-dollar bills folded into bundles with elastic bands, and you were invited to take as much as you liked. How much would you take? And how does that make you feel?

If it doesn't feel okay (after all, you did have permission to take it), why is that? Do you need to give yourself permission to receive?

Were you taught it was good manners to order the cheapest meal off the menu?

Is making money in conflict with your sense of spirituality?

> *It's kind of a spiritual snobbery that makes people think they can be happy without money.*
>
> —Albert Camus

As a child, were you reminded of the war years or the starving children in India? Did anyone ever make it clear to you how feeling guilty or "being greedy" can relieve the pain of the war years, or fix the problem for the starving, ten thousand miles away?

Was "greedy" ever defined?

Do you have any other feelings that undermine you? What about fear? Not the white-knuckled sort, but the type that makes you anxious and uneasy in case you might succeed? The truth is that we really can be afraid of success, and we can see it played out sometimes in sports, when a golfer has a two-inch putt to win the match, and unbelievably, he misses it; sports journalists call them chokers.

An extremely talented graphic designer told me she was scared to mail out a promotional brochure. Though this is something she does for clients every day, she felt she couldn't do it for her own business, because she would never be able to produce it to the standard she would like. In other words because she feared she might not do it perfectly, she wouldn't take the risk of doing it at all.

Perhaps, like the graphic designer, you have a self-defeating fear that you might become visible. By trying to keep yourself under the radar so people don't notice you, you take it a step further by not giving others the chance to reject you. There's no need to; you've already done that yourself.

Here are some scary things that might happen when you set out to charge what you are worth:

- You might draw attention to yourself. If you've ever been bullied or abused, you might fear this will make you a target again.

- You might get called greedy.
- You might start to earn enough so you can quit struggling. But then what would you do and who would you be?
- You might get back your power—but there's the worry. Perhaps you can't be trusted with it; it might corrupt you; you may use it to do harm.
- You might make people jealous of you. Your success could ostracize your family and friends; they might think you've become a different person.
- You might have the chance to live your dreams—but are you ready for that? Dreams are rarely the same in reality. Maybe you'd prefer to live your life wanting, because then at least you keep your dreams.
- You might change. You'll become a tyrant, someone not to be trusted, someone you don't recognize, really horrible, eeeeek…

Is it because your thoughts are faulty?

Life consists of what a man is thinking of all day.

—Ralph Waldo Emerson

Is your inability to value yourself the fault of someone else, or are you willing to take responsibility for your own life? Take stock of yourself now and realize you are not the victim in your life—you are playing the star role!

Be aware of times you are deceiving yourself. For example, have you ever thought: "If I keep my prices down, I'll get the work, because no one else will do it cheaper"? Or: "I'm only doing it cheap the first time; then, when I've shown them how good I am, I'll ask them for more next time." Is that really the truth, or are you kidding yourself?

Are there other ways your thoughts are betraying you? For example, do you not charge what you're worth because you associate some negative things with having too much money? That's not as silly as it first sounds. Think about it: there could be quite a few negative aspects to having quite a lot of money. For example, you might feel a responsibility not to squander it; you might even see it as a worry or a nuisance. Having a load of money might be threatening if others perceive you as a benefactor and you don't have the skills to tell them how to go away.

You could feel guilty, or worried that if you had too much money, you might lose your appreciation of its value; you might lose sight of reality, or lose your empathy for others. You may be afraid that you'll lose the creativity that comes from finding solutions within limits, or you could feel isolated because now you have no need to ask for help.

Are you identifying too much with your client?

"The client is always right." Always? Oh, puh-lease. If you believe the client is more right than you are, you're going to end up believing them when they tell you you're not worth it…yet didn't they come to you for help to solve their problems? Never lower your price because you are afraid of their disapproval. You may negotiate your price for a different reason, but first, collaborate with your client as equals for the best solution.

Here's another classic (and one I've heard often), from clients working for an airline, bank, or multinational pharmaceutical company—this one knows no industry boundaries: "I can't afford it." "I don't have any money."

Me (not out loud): "So why, if you've got no money, are you telling me your problems?" Or: "What? Your shares are trading for US $75 on the New York Stock Exchange, the company made $3 billion profit last year, and you're trying to save ten percent off my invoice for $5,000?!"

My advice: Keep your objectivity and don't buy the client's sob story. If it's really important to them, they'll find the budget from somewhere.

Are you not charging your worth because your client is a friend?

I've fallen into this trap plenty of times. Heck, if you're creative enough, you can rationalize anything! But answer this: Who are you in business for? If you don't make enough money, what else are you going to live on?

Don't be influenced by other people's opinions. Don't let them impose their price on you. You are in business for you. That doesn't exclude your option of making a gift when you want to, but it should always be of your own volition.

Think about the concept of "mates' rates"—is that really what a good mate asks of you? Wouldn't a good friend appreciate your value and be willing to respect that in the customary way? Inevitably there's an imbalance set up when a job is done at less than market rates; there's an unspoken situation of indebtedness, a lurking potential "quid pro quo" that could be called into account at any time...not to mention the risk of hurt feelings and friendships lost.

A marketing expert, Sally, was taking a sabbatical when a family friend, a successful landscaper, called her to ask her a favor. He had employed a marketing consultant who had failed to perform, and he was left in the lurch just prior to Christmas. Sally came to his rescue and spent an entire weekend doing an extensive repair job, for which he was very grateful—but money and payment were never discussed. When Sally told me how she had missed out on time with her husband and children, I tried to lighten things up by suggesting the money might be useful for buying extra presents. Sally told me that though her client probably expected to be invoiced, she wasn't going to charge him—her husband said he didn't think she should.

"Why?" I asked.

Sally didn't quite know why, but there were some vague sentiments that included "because the client is a friend" and "because he knows I'm on leave—I wasn't meant to be working."

"Okay," I said. "But if the boot were on the other foot, would you ask him to drop what he was doing and come over on a weekend to put up a free fence of equivalent value?" I asked.

"Of course not," she said.

"But in a sense, isn't that what you've done for him?"

Reluctantly she agreed that it was. I then suggested that Sally come over and spend the next weekend doing my work. She declined. "Why not?" I asked. "Aren't you as good a friend to me as you are to him?"

She laughed awkwardly.

"Come on. You said yes to him; why won't you help me? Besides, you obviously work for the feel-good factor of rescuing someone. I'm desperate! Please help me!" I wailed.

"Okay, I get your point," she said, then billed him.

She received a couple of thousand dollars for her efforts, there's no change to their friendship, and six months later, she was asked to work with him on another job. Don't you just love happy endings?

Now imagine if Sally *hadn't* invoiced her friend. Do you think he would eventually wonder why he hadn't been sent a bill? How might he feel asking her to do more work in the future? What if he then expected that it too would be done for nothing?

If you provide something for nothing, think about the consequences. What message does it communicate? Does it show you don't care for money, that money isn't your language? Or are you doing it to make people like you? And if you are, is that a direct and honest approach?

Would you be setting a precedent, teaching people how to treat you (like a doormat) or setting a new low that will in some way come back to haunt you (like a friend tells a friend, who tells the entire neighborhood)?

Be aware that in some situations, it could even provoke suspicion because we all know you don't really get something for nothing.

Does your client expect to get it for nothing?

When it comes to bills, no one likes to end up paying more than planned. If a restaurant makes it clear on the menu that there is a corkage surcharge on the wine, or that the vegetables are extra, I accept those terms. However, if at the end of the meal I am given a bill that includes a whole lot of unexpected extras, I feel cheated and would probably not return to that restaurant.

Customers never like to feel that a business is trying to sneak one past them. If there is any doubt, be up front about what clients must pay for. How this is handled can be the cause of much good or ill will.

Save yourself embarrassing misunderstandings with people and think about things from your customers' point of view. Is there anything they might expect to get for nothing?

I inadvertently fell into this trap when I asked Sally (from the preceding anecdote) to read this book.

"Okay, smarty-pants," she said, handing me back my manuscript. "So where do you draw the line? When can you do something for a friend for nothing, just because you want to?"

"That's for each of us to decide," I said. "We all have to set our own compass."

She smiled. "Okay, so let's get personal. What about you and me?"

"What do you mean?" I asked, starting to feel defensive.

"Well, you got me to read your manuscript. Isn't that what I do for work all the time? I mean, would you have asked me to read it if I were an assembly-line worker?"

I could see where she was heading, and she had a point.

I had two reasons for asking her to read my book. First, we had talked about this topic and she had shown interest in reading it. Second, I valued her opinion, her business experience, and her sharp mind—yet (apart from a free meal and a gift) I wasn't planning on paying her for it!

Why ever not? The answer was incredibly insightful. It never occurred to me that she would see reading my book as work. I hadn't put myself in her shoes. I realized afterward that I should have raised this issue with her and discussed terms before I gave her the manuscript to read. I should have given her the choice of whether to give or to sell me her time and expertise.

Later, we discussed how we would both hate to live in a world where no one did something for nothing anymore, but the fact remains that business and personal life boundaries do sometimes blur. We decided that the best policy when asking a friend to help is to make a casual offer from the outset *(e.g.,* over a cup of coffee or lunch). Make it clear if you don't plan to pay, and at the same time, give her the opportunity to back out gracefully. This way you'll both know where you stand.

Is your client a charity?

Like the airplane pilot who tells you to put on your oxygen mask before you try to help others, I advise that if you can't afford to be charitable, you should simply say no.

Don't get me wrong: I absolutely believe in creating goodwill and giving before you receive—but only when it is done from the heart, unconditionally.

Remember also that most charities are businesses, and as such, they should be selecting the best provider for the job and not just the one they can get the most cheaply. Take a look at the fine print in the deals between the major corporate sponsors of the biggest charities, and you'll find that very little is actually being donated for nothing; it's all part of a bigger game of barter.

If you do agree to discount or donate your work to a charity, check your motive first. Do you feel obliged to help, or is it because you genuinely want to? Might it be better to be paid what you're worth and then donate back however much you want?

At the very least, send in an invoice acknowledging the dollar value of your contribution, then write in the actual amount charged, highlighting the discount as a donation. Explain why you are waiving your fees, and offer your good wishes.

Do you lack confidence?

There are brilliant musicians playing unnoticed at cheap gigs or on street corners collecting coins in a cap, some of whom are doing it "to get by."

Despite their talent, somehow they've lost the plot. Any bystander can see that they should be doing better for themselves, but the musician can't see it because somewhere along the way, he or she agreed to settle for less. If this is you or someone you know, look for the pay-off. Are you a true victim of circumstance and unable to stand up for yourself, or are you a victim of your own thinking, having given up on your self-belief altogether?

In a less extreme example, I have an artist friend who had had some bad business experiences and had lost a lot of confidence. Subsequently, instead of asking for a dollar amount, she would ask people to pay her what they thought it was worth. She did this even though she was always disappointed; no one ever paid her more than she hoped for. Another friend had a favorite line: "Make me an offer I can't refuse." Not once did he get an offer he couldn't refuse. When we discussed this in greater detail, he realized that he was giving away his power—his right to ask for what he was worth—and he has since learned to be more direct and simply ask for what he is worth.

Are you always giving away discounts?

Discounting, as a business practice, is so entrenched that I probably don't need to help you recreate the arguments that justify it.

The dictionary describes the effect well: "to deduct from the amount, cost; to allow for exaggeration; to disregard; to make less effective by anticipation."

Be mindful when you're offering a discount. Why are you doing it? Is it an investment, and will it net you a greater financial return in the future? Or is it something you do all the time, a thinly disguised (yet noble!) excuse for not charging your worth?

Are you offering a discount as the "lazy way out"—instead of making the effort to explain or demonstrate your value? Before you discount, stop and think: is this the only way to give value?

Do you not charge enough because you are not being businesslike?

If the terms of the transaction are vague, you leave yourself open (and vulnerable) to the other person's generosity.

A few years ago, I took on more work than I could handle, so I decided to subcontract some out. The person I normally used was unavailable, but since the job wasn't overly technical, I decided to try out a relatively inexperienced new graduate (a friend of a friend).

"Thank you. Thank you," said Allan, pleased to get a "foot in the door" and recognizing that the finished job would look good in his portfolio.

"Of course, I expect to pay you," I said. "How much will you charge me for this?"

"Oh, you don't need to worry about that," he said, rushing his words, sounding embarrassed.

"I insist. What do you normally charge? You must have a rate, surely?"

"Oh, we'll work it out later," he said.

I left the project with him, but how do you think I was feeling? Though I had no doubt we would "work it out later," I felt very unsure about this. My imagination went into overdrive: What if he overvalues the job and wants to charge me ten times more than I would normally pay? Then worry kicked in. It could get complicated (remember, I don't like conflict), so I started to think that perhaps I should cancel him and wait for my usual copywriter to return.

Next, I thought about the situation he had put himself in—how he was the one who should be worried. Not only did he risk me taking my business elsewhere—not an outcome he wanted—but, because he had indicated that he was prepared to do the work for nothing, he was at the mercy of my generosity.

Do you see how customers need to feel confident when they place their business with you? You are obliged to be businesslike; it's good manners and good business.

Do you underestimate the amount of work involved?

Do you undercharge by underestimating the amount of work involved? Is that because you are an optimist and you always think things will go smoothly, or do you need to keep a better record of time and costs involved?

Another way you can wind up subsidizing the customer is by not adequately covering costs when there are changes to the specifications of the job. If you are doing this, you may need to seek the services of an accountant or business coach to help you put better systems in place. In the meantime, keep reading this book. Even with good systems in place, some people still make this mistake because of a faulty belief system (it's not worth bothering about), or laziness (it's too difficult). If the job changes, you must make provision for this and re-quote, and be clear in your communications.

Are you worried about the consequences?

Here's a thought: What if you and all your colleagues charged $200 an hour, then someone lifted their rate to $300 an hour…wouldn't you want to raise your price too? Until it happens the next time, the time after that, and so on—you might then be afraid that if everyone did this, prices would go up and up and you'd never be able to afford what you wanted.

We all have a vested interest in maintaining the status quo.

If you decide to put up your prices, you will probably touch a nerve in price-sensitive customers. But you can still do it. Just be aware that your customer needs reassurance that you're not going to "rip them off." They need to know they can still trust you and that they'll still receive the quality they expect. Be sure you deliver on your promises.

Do you think you don't have enough time?

If you are not charging what you are worth because you think you don't have time, I feel complimented that you are reading this book. I also hope you will soon be motivated to find the time, because if, in your business, you are just "doing" instead of managing what you "do," you'll remain a mouse on a wheel getting nowhere. There will always be a hundred and one reasons for not addressing this issue; it's up to you to find your own internal motivation, and it's time to make yourself your top priority. Stop thinking you are not worth it, and believe that you are.

CHAPTER 2

UNDERSTAND YOUR RELATIONSHIP TO MONEY

If you've ever done something or gone somewhere you didn't want to, just because you've paid for the tickets and couldn't get your money back, then you have been a slave to money.

What is your relationship to money? Is money your comfort, your god, your friend, your master, servant, lover?

In a sense, money does "talk."

In English, Japanese, Taiwanese, or French, two simple words ("How much?") and an open wallet can get you around most of the world. In a capitalist system, we need money to function, and a big part of you is the way you handle, control, manage, lose, fritter, invest, eat, burn, love, hate, or worry about money.

The things money can buy have probably defined your experience of holidays, birthdays, and Christmas—and some of your most deep-seated values. For example, were you brought up to "get your money's worth"? What happens now when you don't? Do you end up feeling cheated or "ripped off"?

Think about the things money symbolizes to you. When you were a child, what were the conditions of pocket money? Did money bring you joy and happiness, love, entrapment, resentment, or fear?

As an adult, what is your definition of waste or extravagance? I have friends at either end of the scale when it comes to grocery shopping. One buys a lot of sausages and cheap mince and prides herself on her economy; the other spares no expense and buys exotic fruit, fresh salmon, and expensive, lean cuts of meats without exception. Her argument is that you can buy a lot of quality food for the price of a triple heart bypass or a mobility scooter!

What does prosperity mean to you? Some financial advisers advocate that you save $3.50 a day (the cost of a cup of coffee) so you can reap the benefits of compounding interest and retire in moderation years later. I was inclined to agree with this advice until the day I realized that having the disposable cash and time to enjoy a bought coffee a day *was* prosperity. It was neither a wasted opportunity to save nor an extravagance.

Money means different things to different people, and it can buy us experiences that are unique to us.

A friend of mine told me her dream was to buy a brand new Porsche. Bridget had worked out that she could afford it if she added the loan to her mortgage and paid it off over twenty-five years. Being financially savvy, she knew the real cost of the car, but said it was something she just wanted to do in her lifetime, so the expense would be worth it. When I found out she hadn't yet driven one, we arranged a test drive. We had only been driving five minutes when I asked her if the car "did it" for her, and was it worth it, to which she replied, "I don't know. I think I might sooner have six months skiing in Aspen."

We discussed how she would feel returning to the workplace in order to pay for it. She told me she wouldn't have a problem owning a better car than the general manager, but she would find it difficult going back to the boring job she had. To her, that car was a metaphor for the excitement that she otherwise lacked in her life. Buying it would have provided the biggest adrenalin rush; after that, it would have been downhill all the way. What she really wanted to do was break out and test her self-belief. Fortunately, she realized in time that a car repayment plan wasn't the answer.

It takes one to know one: think about your own style

Some cultures teach their children to barter and bargain. From my family upbringing, I always knew a price was negotiable. If you found a fault in the merchandise (and you always looked for one), you would ask the salesperson to knock off some money. Other people might go shopping and never dream of questioning the price.

In short, the buyer has a style, as does the seller. Some shops cultivate a style deliberately—think about the very rich shops, where you know by the response that "if Madam (sniff) has to ask the price, she obviously can't afford it!"

Here's another example: My husband and I bought the land we live on at an auction. I had been taught to dress down *(i.e.,* don't look too affluent) when going to buy something expensive; that way, the seller won't hope for too much money. Before going to the auction, we had a strategy and definite price point. We dressed inconspicuously, blended into the crowd, and didn't join the bidding until it had started to slow. Soon the bidding tailed off, and as it wasn't at its reserve, we ended up with the last bid and the right to negotiate with the vendors after the auction finished. We eventually bought the site, paying just $500 more than our very modest budget.

Several years later, I was talking to a woman who bought her house at an auction, and I asked about her approach. She told me she got dressed up in her most expensive black leather suit and gold jewelry. The auctioneer asked for bids, and as soon as someone started, she launched in, jabbing her hand in the air immediately when anyone else raised the bid. Her strategy was to psychologically overpower the other bidders by making them think she was out to buy the property at any price; they might as well give up early and go home. She got the house she wanted at what she thought was a bargain price.

By comparison, a large sub-divisible plot of land was put up for auction a short drive from where I live. The land was expected to earn somewhere between three and four million. After development costs, it was certain to return a developer some healthy profits. A small crowd turned up on the day, and the auctioneer gave his preamble before asking for an opening bid. "Five million," called out an agent working on behalf of an out-of-town client, and that was it. Sold! According to the grapevine, the purchaser knew he could make a profit at $5 million and was happy to pay it.

How closely do you identify with money?

A friend's partner loved the type of work he did, but resigned from his job out of desperation because of the internal politics in the office. For a while, Gavin got seriously glum as he considered other options before being wooed back to his original employer as a consultant. Though he was doing the same work, he was on a six-month contract, at an hourly rate that was twice his previous salary. When that contract expired, it was renewed at almost double the rate again! My friend was laughing as she told me how Gavin was a changed man. The job was the same, and the politics remained; the only difference was that he was now being paid more than a hundred dollars an hour.

According to my friend, it was *because* Gavin was paid more that he felt more valued, which raised his self-esteem. External events changed his internal experience. Other people might work in reverse: they may need to raise their self-esteem first before they are able to receive more money.

I know a woman whose husband died in an air crash, and she received a large life-insurance payout. Her problem was that since she was publicly known to be cash-endowed, she was never sure whether her new male friends were interested in her or her money.

I also know a man with the opposite problem. In his words: "No one will have me because I've got no money and no prospects."

Though two extremes, the problem for these people is the same: they both mistakenly believe that they are their money. They possibly also believe that everyone else shares the same money obsession they do, when they might not.

To make this point a different way, consider the last time you saw someone you paid to do a job for you. When you last bumped into her at the shops, was your first thought: *Oh, there's Debbie, my cleaning lady whom I pay $18 per hour,* or did you think: *There's Debbie; doesn't she look great in those fluorescent tracksuit pants?*

Similarly, can you see that your own value is far greater than any given amount of money? The last time you saw a customer in a supermarket aisle, do you think they thought: *Oh, there's thingamee, and she charged me $180.* Somehow I doubt it.

Don't identify with money. When you take the emotion out, money is just a number, no more a measure of your character than your height is.

Are you master of your money, or is it master of you?

> *If money be not thy servant, it will be thy master.*
> —Sir Francis Bacon

Charging what you are worth is going to be difficult if you have a negative block about money.

What is the most money you have ever had as savings?

How high could the balance get before you reached a point where you had to do something like spend, invest, or gamble it? Ten dollars, one hundred, one thousand, one million, ten million…?

Think back over your lifetime to see if you have a ceiling in terms of salary, sales, turnover, or profits. Do you have a psychological earnings set-point you can't seem to go beyond?

Conversely, do you have a debt set-point—a level at which you are jolted into action to exercise an internal "credit limit"?

Realize that you put up your own blocks that may work against you. Some are simply bad habits that you can change, but others may be attitudes that are active subconsciously, sabotaging your best interests.

You need to be fully and consciously in control of your money, or you could find the emotion of the numbers affecting your objectivity.

Will you ever have enough?

In my opinion, there is something very sad about the "affluent poor"—the people who, no matter how much money they have, are unable to enjoy their wealth because they still act poor.

How much money will make you feel satisfied you have enough?

If you don't know, or trying to decide makes you feel fearful, take a look at the way the financial industry promotes itself. Notice how the experts instill the fear that you will never have enough. Observe how you're being marketed to, and remember that marketing is about creating demand, not satisfying it.

So take a deep breath and decide how much is enough—how much you need to earn versus how much you think you should be earning. Is it possible that you are feeling pressure to charge more than what you do, when what you are earning now is just about right?

Money and opinions—yours and mine

> *I've spent ninety percent of my money on women and booze; the rest I just wasted.*
>
> —George Best, British soccer star

Money can evoke the full gamut of emotions—fear, guilt, envy, love, greed, sorrow, joy, pride, justice.

People are very quick to make money judgments, some of which are very personal. I recently saw a woman's magazine advertising its lead story: *Named and Shamed! The stingiest stars in the planet.* Inside, the article is headlined: "Famous and frugal…Miserly millionaires who pinch pennies despite their bulky bank accounts."

In my own experience, a dinner party can be divided into warring factions in no time at all by simply raising the question of whether an interior designer should charge a fee for her advice *and* receive a kickback from the companies from which her clients purchase goods.

On one side, there will be those who think her fee should be enough and that her advice should be free from commercial bias. Then there are those that believe that if she can get the extra commission, good luck to her.

Ultimately, so long as you don't break the law, you are entitled to your opinions, the same way I am entitled to mine.

Are you conflicted about profit?

Do you know other businesspeople making huge profits? What words do you use to describe them? Do you call them successful...or greedy, dishonest, criminal, immoral, or something else?

If you don't have a high regard for others when they are making good money, how will you have a high opinion of yourself when you're doing the same?

A while ago, I read a newspaper article about Hollywood hairdressers who differentiate themselves by calling themselves "stylists to the stars." Apparently some earn up to $8,000 plus expenses every time they run a comb through an A-list actor's locks. They see their service as good value for money.

Think about your situation. If your service and products meet the legal standard, and your customer is happy to pay you, why wouldn't you charge magnificently?

Returning to Hollywood for a moment, the studio accountant who objected to paying the bills (that included first-class airline tickets, limousines, and rooms at the best hotels for the stylist's entourage) said, "It's out of control. It's mind-boggling what these people get to blow-dry hair. They are treated as if they are the talent."

One stylist replied, "Hair can be the first thing you notice about a person...[and later, about her prices] I am not holding a gun to anyone's head." Another said she thought the studio heads were "lashing out at us to protect their own corporate jets. The stars will protect us—after all, they don't get beautiful by themselves. You should see some of them in the morning. That would sell more tickets than some of their movies." *(The Sunday Star Times,* November 30, 2003.)

Is there a psychological price barrier to the level of profit you or your customers believe you should be making? Is there a level of profit that, if you exceeded it, would make you think you had made "too much"? Who's to say what that limit is?

How much profit do you need? How much profit do you want? How high does profit have to go before it's immoral? Is it more moral to price your services so low that you make losses and have to lay off staff?

In a capitalist culture, so long as you are telling the truth, no one is harmed, and everyone is happy with the deal, you need never feel apologetic for how much profit you make.

Is it okay to make $200,000 in one week on a property deal, but not okay to make the same on your labor?

Do you believe your profits have to be "hard-earned"? I know a couple who had planned their get-rich path and were making headway, when they were given $80,000 by a parent. For a while they felt "pipped at the post," until they opened their minds. *Then* they went out and celebrated!

Look closely at the answers to your questions. Do you see where you hold conflicting ideas? For example, you'd like to earn a million dollars in your spare time, just like Barry Jones, but you don't want to lie like he does. Or you want to make a 300 percent markup on an item you make, but you think that's being greedy.

Work out what your values are, and be aware that you will never feel okay about charging what you are worth if you have some internal conflict. Only you can align your beliefs with your behavior; no one else can do that for you.

The illusion of rationality

Princeton University cognitive psychologist Daniel Kahneman and Vernon Smith, professor of economics and law at George Mason University, won a 2002 Nobel economic prize for designing experiments to help understand why people make irrational economic choices in a theoretically "rational" marketplace.

This type of economics is called "behavioral finance," and it studies the inconsistencies between the logical thing to do (financially speaking) and what we actually do. For example, have you ever spent ten dollars driving to the other side of town to save three dollars on an item you could have gotten from the corner shop?

Let's pretend I was going to give you a gift of money. Would you rather take the $1,000 I offer you today or $1,200 in a couple of months?

Researchers have shown that something as small as a few sweets can affect the behavior of a group of people when deciding the answer to that question. Those who were given the sweets before they were asked the question were quite happy to wait a couple of months, whereas the ones who had to answer the question before eating the sweets chose to take the money now.

We like to think we are rational, but the fact is that most people don't save enough for their retirement. This phenomenon is known as "future discounting," wherein we value money in the here and now more than money twenty or thirty years later.

Likewise, though we like to think we learn from past mistakes, we often don't. We tend to be masters of self-deception, and there are many ways in which we actually hinder our learning. Economists have identified and given labels to some of the ways we distort the truth. See how many you recognize from the following:

- **Self-attribution bias.** This is where we tend to attribute good outcomes to our skill, and bad outcomes to the luck of the draw. With this bias, we cannot learn from our mistakes, simply because we don't see them as our mistakes.
- **Self-serving bias.** This is where people willingly take credit for success, but will not accept the blame for mistakes or failure.
- **Overoptimism and overconfidence.** Our trust in our own opinions is greater than our trust in the opinions of others. A teacher asked a class of students to put up their hands if they thought they would score in the top fifty percent of the marks in a test. Eighty percent raised their hands.
- **Conservatism bias.** Have you ever found it hard to change your mind? People often have difficulty going back on a decision, especially if it disagrees with the status quo.
- **Confirmatory bias,** where we look for information that agrees with us. Think how good it makes you feel when talking with people who share the same view of the world that you do.

- **Hindsight bias,** where we like to think we would have predicted what happened before it happened. The indicators appear to be loud and clear *after* the event.
- **Availability bias,** where the most recent (or most memorable) information influences our objective judgment. For example, your neighbor gets assaulted and robbed, and that makes you feel your locality is far more dangerous than statistical evidence suggests.
- **Loss aversion.** Share market investors often show more interest in avoiding a loss than in making a gain, resisting selling shares that are going downhill.
- **The "persuasion effect,"** where investors are more likely to be influenced in their decisions by a credible source rather than a credible argument. (This is the reason why testimonials work!)
- **The "illusion of knowledge."** This is evidenced in those who believe the accuracy of their forecasts improves with increasing amounts of information. (It is the quality of the information and what you do with it that counts.)
- **The "illusion of control."** This is the belief that we can somehow influence uncontrollable events, such as the roll of the dice or the winning lottery numbers.
- **Representativeness.** Our minds like to take shortcuts by sorting information into patterns, for familiarity. Sometimes we jump to the wrong conclusions because we judge events by how they appear, rather than by how probable they are.

Source: Shefrin, Hersh. *Beyond Fear and Greed: Understanding Behavioral Finance and the Psychology of Investing* (Oxford University Press, 2002).

Chapter 3

Know Your Client

Emotions have the capacity to turn us into virtually different people—spendthrifts or tightwads.

—Professor George Loewenstein, Carnegie Mellon University

The previous chapter focused on us getting clear about our own relationship to money. Now let's turn our attention to understanding our customers.

As frustrating as it can be, clients are often unpredictable and illogical, but they have a perfect right to be this way. Sometimes they are simply too stressed, too hungry, or too tired to make a rational decision.

Clients are just like us: they buy to satisfy an emotion too.

They like to believe they are right all the time, yet they make mistakes when they make decisions—just like we do.

What's more, they'll probably only hear half of what you say and will remember a third of that—just like we do.

And more often than not, they'll like you more if they agree with what you're saying—and they may not listen to you if they don't.

What do you know about your client's money mentality?

The better you understand your client's money mentality, the greater success you'll have in doing business with them. Unless you are the government or have a monopoly on an essential item, you simply cannot make clients spend money unless they want to.

Some clients will always want the cheapest thing on offer, while others will choose the most expensive. Your job is to convince them you are good value according to their price mind-set.

Never transfer your money mentality onto others; the customer is different than you.

The person who is used to spending ninety-five cents on a ballpoint pen is going to need extra help spending $49.95. Conversely, what you might think is expensive could be trivial to your customer. A good example of this occurs in the movie *Pretty Woman*. A millionaire (Richard Gere) is attracted to a prostitute (Julia Roberts). When he asks how much she would charge to stay with him for the whole week, she bids him up to three thousand dollars. Later, she tells him she would have accepted two thousand, to which he quickly replies that he happily would have paid four.

When you're talking with your customer, see if you can sense their emotional intelligence around money. Are they "penny wise and pound foolish"? Are they thinking short term or long term?

Look closely at your client. How does your client "tell on" himself or herself? Is it by their body language, their shoes, the clothes they wear, their haircut, the car they drive, the brands they identify with, their accent, manicured or bitten fingernails, or the qualifications on the business card?

Though we may make a snap judgment when we first meet someone, that impression may not be accurate when it comes to forming an opinion about their money mentality. Beware: the prestigious car, the elite watch, and the "in" T-shirt and sunglasses may all be disguising huge insecurities and paid for with borrowed money.

I've found that actions speak louder than words, and I'm more inclined to "test" people for their money attitudes as we talk. For example, I'm often fascinated by the dynamics of who pays for lunch. Who wants to play the parent and buy it for everyone? Who is the most reluctant to reach for their wallet?

Resolving the seemingly trivial issue of who pays for lunch can reveal so much about the other person, the level of communication, and the power dynamics within the relationship.

Observe how it feels when someone buys you lunch. Do you feel disempowered, loved, appreciated, or "bought"? Is that the way you always feel (because it's one of your own hang-ups), or is it saying something about that person?

What clients say and what they really mean

Sometimes what clients say and what they really mean are two different things. Often, despite what they say, it is not the price that's holding up business.

Clients don't tell the truth *not* because they're compulsive liars with character flaws, but because of the way they've been socialized. The cost is often used as a convenient reason for backing out of the deal. "I can't afford it" may be short for "I would buy it if you changed this and this, but I can't be bothered explaining all that to you when I can just go to XYZ and get exactly what I want without having to go through all of this."

Price is often a mere excuse, albeit effective, because we know that if the seller has a problem with this issue, he or she will wilt away, easily intimidated.

Selling is not just about whether or not your client has the money; it's about whether they spend it with you. This is the very reason you need to learn how to negotiate: to find out what they really want, and whether you can provide it while making a profit at the same time.

In real life, your client probably says things like: "I have to think about it." "I'm not sure about it." "I'm waiting for a quote from another company." "ABC will do it for less." "I didn't want to spend that much." "I've changed my mind." "Someone else told me about another company they had used." "I got a better one from XYZ; my accountant has dealt with them before."

Clients have their own money issues and hang-ups, which may include:

- Feeling scared to commit
- Needing to get a bargain or "something for nothing"
- Being worried they'll be ripped off
- Feeling inferior because their budget is too small
- Being kind—they don't want to hurt your feelings
- Being unassertive or indecisive—they simply can't decide or say no to anyone
- Being reluctant to tell the truth; *e.g.,* they don't want to admit they are just doing their research, or they already know who they're going to use
- They're following someone else's orders and don't actually have authority to buy

Educate your customers

Most people won't willingly release their money without some form of resistance, so part of your job is to make them feel good about it. You need to develop their trust and put their minds to rest that you are the person who can make things better for them.

You know all about your products and services, but your clients will not. There may be huge gaps in their understanding of what you do and how long it takes. They may be ignorant of the knowledge and skill required to do what you do, and what's more, they do not know what they do not know. They may not see the value in what you provide, so it is your job to educate them.

Can you demonstrate your product? How many of the senses can you involve? Can you prove your cost-effectiveness? Do you have a graph to show savings? What about case studies, before-and-after pictures, or testimonials from happy customers?

Teachers of fiction writing repeatedly tell their students: "Show; don't tell." In other words, don't tell the reader the soldier went to war and felt miserable. Show the reader his misery—the rats in the trench, the incessant rain, his cold hands against the steely bayonet, and the mud that made the flesh around his fingernails rot like…you get the picture!

Show your clients the advantages. Show them what it means to them—how it will save them time, make them more money, make them thinner, younger, more gorgeous.

For example, if you are selling tax advice, show your customer a testimonial from a happy client who saved $5,000 in taxes last year after spending just $400 with you.

Now imagine you're selling a pre-cracked fresh-egg mix to a meringue company; work out the savings in breakages, transportation, and packing costs of a liquid delivered in a bucket rather than eggs in cartons. Work out the labor savings in egg-cracking time, and add up all these savings in a week, a month, or a year. Now show your client a picture of the increase in meringue sales and a graph of the extra profits, and how one cent more per meringue converts to $100,000 in extra sales per annum. Whip up a batch of fresh meringues and let them taste the difference.

Provide excellent value

> *Give everyone more in use value than you take from them in cash value.*
>
> —W. D. Wattles

Value is in the eye of both the seller and the buyer. The disparity in value is never more apparent than when you go house hunting—the vendor will always see far greater value in their castle than you do.

Value is a complex concept that is relative to the customer, the marketplace, the timing, the problem, and the product or service.

If you know very little about your customer, you cannot know whether they see your price as representing good value. Your aim is to provide the customers with what they perceive as good value, at a price that returns you a good profit.

Interpret your customer's desire for "more prettier"

> *All the wodes [roads] should be pink, and if they were sparkly, they would be more prettier.*
>
> —Brittany, four years old

To stay in business, you need to understand the needs of your customers. If your customers are four-year-old girls, how can you make things more pink and sparkly?

There are three things I've found that customers like, crossing all cultures, ages, interests, and socioeconomic boundaries:

1. People like to get their money's worth and feel like they are getting something for nothing.

    I still support the jeweler who estimated $230–$240 to repair a necklace, but when I went to pay, he only charged $200. I like generosity, and I value a display of honesty.

2. People like to be made to feel special.

    Remember your customer's name, and give her a lollipop or a pink, sparkly sticker if she is only four years old. These things cost so little, but business is about people, not just money.

3. People like to deal with people they know and trust.

    Familiarity is the number-one reason for using a product or company. Word-of-mouth advertising is the cheapest around. When you finish a job, call back later and check if they're happy with what you did. It's called follow-up, and it is a great way to check whether your clients are satisfied or not. If appropriate, ask for a testimonial and see if they're willing to give you the names of others who might want what you have just sold them.

What sort of clients do you want to attract?

The goal of your marketing and advertising efforts should be to attract the type of clients you want. My favorite clients hang on every word I say, love all my ideas, give me exciting projects with huge budgets, and pay me handsomely and promptly…and these are the type of clients I want more of.

Are you dealing with "your type" of customer? If not, ask yourself why not. Identify common characteristics of your best customers, to help you recognize and search for "clones" of clients like them.

Be aware that if you price yourself too low, you may even attract clients looking to cut corners, clients with poor credit ratings, or clients who simply can't afford to pay the going rate. This can result in an erosion of self-esteem, which makes you feel negative, bitter, unhappy, stressed, or resentful; which in turn may detract from the quality of your work. Things may start to go wrong (a by-product of cutting corners or working to exhaustion, and sometimes of having a tough customer), which erodes reputation and self-esteem for the next job, and so on. Plus, be seen to be too cheap, and you'll annoy the good operators in your industry, who will actively assist your descent on the downward spiral.

Not all customers are created equal

John is an artist who has two distinct categories of clients: those from whom he makes a living and those he targets for "showcase" creative opportunities. Often the clients who are prepared to take the creative risks are smaller companies with more modest budgets, but he takes them on, even though his return on effort is less than he would like. In effect, he is subsidizing their work, but in his view, it doesn't matter—he sees it as marketing, because he ends up with something to put in his portfolio to attract the better-paying (but less interesting) "bread and butter" clients. Looking back over his last years' time sheets, he realized that while only five percent of his clients were in the showcase category, they took up twenty percent of his time, making them four times less profitable than other clients. To make sure he doesn't lose sight of his financial objectives, he carefully monitors how many of these clients he takes on.

Like John, you might be having the most fun with the least profitable clients, but the bottom line is this: If you spend all your time with clients who are only willing to pay $10 an hour, when will you have time for those ready to spend $100?

It makes sense to spend your effort attracting customers that generate $100 profit rather than those who make you $2. However, as in John's case, there are sometimes good strategic reasons for taking on those who are less profitable. It may therefore be wise to segment your clients into groups according to profitability first, then into other groups according to the business objectives they meet.

For example, you may group your customers according to frequency of purchase: are they first-time customers, occasional purchasers, or loyal advocates? Alternatively, you may want to group them according to their potential for future business: what might their lifetime value to your business be?

Be mindful that you don't make the mistake of consistently offering new customers better deals than your existing ones. This has been a common practice in the telephone, insurance, and banking industries in particular. Customers aren't stupid, and if they can get a better deal for being a new customer, they will *become* new customers to reap those rewards. The industry term for this is "churn," and many marketing personnel in some of the largest corporations are kept occupied trying to reward customers for loyal behavior and reduce customer churn.

Chapter 4

Know Your Options

Pricing is part psychology, part fiction. It is as much about emotion as it is about numbers.

It is my intention in this chapter to expose you to a wide range of ways to decide your price. Not all will apply to your situation, but read them anyway, to broaden your understanding and flex your mind to the possibilities.

For example, if you are a coach and charge for your advice, you could take a commission. Or charge a flat fee, an hourly rate, take a profit share, or work on a pro bono[1] or pro rata[2] basis. You could charge according to the market rates or the value of the job, or by donation only.

Perception is reality

Which words go together: cheap, long-lasting, nasty, quality, junk, expensive?

> *Quality is remembered long after the price is forgotten.*
> —Gucci family slogan

---

1. Pro bono: done or donated without charge; free. Pro bono legal services are provided free to a client with the proviso that only if the client wins their case, they will pay the agreed fees.
2. Pro rata: In proportion; proportionately; according to the share, interest, or liability of each.

Think about some of the world's most prestigious brands, and you realize that the high price is the point.

There are times when the price is an integral part of the product's success, like some of the stop-smoking or weight-loss programs: if you pay $1000, you'll be a lot more motivated than if you paid $10.

A while back, I paid $170 to a sports laboratory to receive advice I could have read in a book about how to train for a marathon. But it was only because I had paid for the advice that I actually acted on it, and it worked. Instead of wearing myself into a state of fatigue, overtraining six days a week, I did as I was told and trained every second day. Had I done it my way, I would almost certainly have ended up injured well before the day of the marathon.

Sometimes the packaging, or the shop decor, or the advertising creates a higher-quality perception, so you can charge more. Even if the cost of the ingredients was the same, you would still expect to pay more for coffee on the waterfront at Monaco than for coffee in a polystyrene cup from the express bar at the petrol station.

If you are pricing yourself too cheap, you may be jeopardizing business opportunities because customers won't believe you can possibly be as good as you say you are—the price doesn't indicate that. Perception is reality. Just think about the loyal Black Label whisky drinkers who pride themselves on their good taste, but when blindfolded can't differentiate the quality.

What does your price tell the customer about you?

What can your customer afford?

I know some psychologists who charge different hourly rates for customers in different income groups. If you have a household income of $100,000 or more, you pay the highest consultation fees.

In most situations, asking customers to tell you their income would be an unpopular move. It may also be problematic dealing with those who are tempted to be dishonest in order to receive lower fees, but I mention it as an option, to open your mind to alternatives.

If you are working in a lower socioeconomic area, you may need to consider how much your customer can afford, the volume of potential business, and what would be a viable price for you to make a profit. Perhaps you need to offer financing, so the customer can break the purchase into smaller monthly payments. To make it simple for you (and less risky), you could partner with a finance company.

If, however, you are marketing yourself to a wealthier segment of the population, the question is not how much your customers can afford, but how much they are prepared to spend. And do they trust you enough to spend it with you?

Customer segmentation

Customers in socioeconomic area A may only have one quarter the spending power of those in area B, but there may be ten times the number of customers in area A, all of whom are easier to sell to than in area B.

Should you go fishing for the occasional big fish in area B, or head over to the large pool of fish and catch lots of sprats?

If you segment your customers into groups, you do not have to make an either-or decision; you can do both. However, if you are going to offer a more affordable deal to customers in area A, then you may want to alter the product or service in some way so customers in area B don't demand the same low price as their poorer counterparts. This is called product differentiation.

Consider your costs

If you are selling something, you might be concerned about the costs you incur while doing business, but you can be sure your customer *isn't* the slightest bit interested. All your customer wants is a good price.

Costs should not be the sole determinant of pricing. As we shall see in the pages to come, though operating costs need to be factored into your price, there are many other things to consider.

Price with an eye on your competition

Price-wise, where are you ranked in the market? Are you a price-setter or a follower?

Your competitors may set a price range, but where you are placed in the market is your decision.

If you are a new entrant to the market and far superior to the market leader, then you might want to rush out and tell everyone you are the latest and greatest— because *you really are.* You can also charge at the high end of the scale, or set a new high.

When you compare yourself to your competition, you promote them. If you are going to use competitive pricing, compare yourself with what is likely to be the best available, the price leader. But if you don't have a convincing case, don't mention the competition and don't try to compete on price. Instead, offer a better product, a unique service, more choices, nicer staff, or a better guarantee. Be the best you can be at being you.

Market rates

When setting your price, you may be advised to research the "market rate" as a useful reference point, but bear in mind that, official as the term sounds, there is actually no such thing. Market rates are a range. Setting your price according to the market rate is a guessing game, because markets are always dynamic, expanding and contracting according to a wide number of variables. The advent of new products or competitors has an impact on the marketplace, changing the dynamics. New technologies and distribution channels offer the consumer new choices, sometimes influencing or changing consumer needs.

Supply and demand

The market may dictate price when there is an imbalance in supply or demand. If demand weakens or there's no money around, prices collapse. Conversely, if there is strong demand and limited supply, the price moves upward.

If you and ten others are vying for the work, then the buyer can play suppliers off against each other, and you could end up doing the work for nothing…or less. There may be some situations where you decide there's a strategic reason for doing that, but you wouldn't want this to happen by accident.

Price and volume

Traders always try to negotiate more for less. How much for one orange? One dollar. How much for a dozen? Ten dollars.

You too may want to base your prices on the volume of service or products you are able to produce, but first consider this: If the price were lower, would people buy more?

The good news is that if you can make more for less, there is no rule that says you have to charge less. However, when the product you sell is your skill or knowledge packaged as parcels of time, you can't make more time. This topic is covered in greater depth in a later chapter.

Price to suit the changing times

In most markets, just like in nature, there is an evolutionary progression where newer (and supposedly better) services and products replace old ones.

Marketers talk about products going though a life cycle equivalent to birth, childhood and adolescence, middle age, old age, and death. When a product is first launched, there may be a window of opportunity where customers are excited and willing to pay more to have the latest and greatest; during this time, you'll be able to make some headway in recovering research-and-development costs.

Charging a higher price on entry to the market is called "price skimming" in marketing jargon. Price skimming allows businesses to skim off higher profits from customers who are keen to pay a premium price to be one of the first to own a product. If demand for the product is high and supply limited, then skimming allows companies to take advantage of the small window of opportunity before new competitors come into the market.

Lowering your price to appeal to a large proportion of your target customers is called "penetration pricing"; this tactic is often used to gain market share and will sometimes have the effect of driving weaker competitors out of the marketplace altogether.

Later in your business life cycle, when your product or service is seen as yesterday's news, you'll have to think creatively to keep the prices up. It may be necessary to repackage or re-brand what you have to sell so that you appear new, better, different, or unique in some way; once you've increased demand for the new and improved version, see if your price can go up too.

Usage rights

Photographers and photo libraries go to great lengths to protect the usage of the images they supply. Each time the image is used for a different purpose, more royalties are due. Is it possible you have some products or services you could franchise in this way?

I know of one artist who sold his work under the proviso that if his art was resold in the future, then he (or his estate) would be due a commission.

Learn from the alpaca breeders

My children were stroking the nose of an alpaca (looks like a llama, only smaller) at a country show, begging me to buy them one to put in the front garden. I was told the asking price for an alpaca as a pet was $1,000, but if we were planning to breed, a female would cost us $5,000. The female "pet" alpaca would be sterilized to make sure we didn't change our minds later, and male alpacas were not for sale.

Have you figured out a way to charge more for a product with the ability to multiply?

The value of scarcity

Some markets are manipulated to create scarcity, to add value. Think of the huge dollars some magazines pay for an exclusive on a celebrity wedding. Think of diamonds, and limited-edition stamp collections.

You can charge more if you are the best, the fastest, or the most experienced, or if you are very rare or exclusive.

> *A racehorse that can run a mile a few seconds faster is worth twice as much; that little extra proves to be the greatest value.*
>
> —John D. Hess

Put a price on time

Customers are willing to pay more for an aged bottle of wine than this year's vintage. Antiques and the work of deceased artists command escalating amounts.

If you are in a position to save a client time, what is the opportunity cost of that time worth to them? People are willing to pay to reduce the amount of time they might have to spend waiting.

Can you find a way to "manufacture" time? What do you think plant centers are doing when they sell you a five-year-old tree rather than the seedling?

Take a commission or a percentage

Some industries rely on value to set their price, as real-estate agents do when they take a commission based on the value of property sold. Others take a percentage of the value of expenditure. Architects often work this way.

This method of pricing suits some professional associations that have done the spreadsheets and worked out a formula that gives the best return. Generally the percentage is higher on smaller jobs, to protect you from losing money on time-intensive low-value jobs. As the value of the job increases, the percentage decreases or levels off.

Put a price on your brilliance

If your time and knowledge is what you have to sell, you may find there are times that a job has a value well in excess of the time it actually took you. Let's say you are an architect designing a town hall and you create a fantastic monument that can be built in a way that shaves $300,000 off the construction costs, yet it was all in a week's work to you. Don't you think you should charge this client more than next week's client, who wants you to design a garden shed?

Find ways to be creative—maybe invent a "menu" of charges. Turn some of your services into products by naming them and building a sales story around them. Invent categories or classes of service, like bronze, silver, gold, and platinum care, or economy, standard, deluxe. Ask for royalties or a share of profits. Invent a performance bonus scheme whereby the more your client succeeds, the more you do too. If you don't put a price on your experience and intellectual property, no one else will.

Retainer and royalties

There is no such thing as a "sure thing." Until *Lord of the Rings* played in the theater, Peter Jackson did not know for sure that the world was going to love his movies.

Would the client like you to share some of the risk? Could you arrange a retainer to cover overhead, and negotiate a royalty deal if the commercial venture is a success?

Model and celebrity agents often arrange deals for their clients that include payment based on both retainer and royalties. Open-ended royalty agreements ensure that the harder the celebrity works for the company or product they endorse, the greater the returns they receive.

Your client will be glad to pay for your experience

Clients *expect* to pay more for a lawyer with twenty years of experience.

Imagine the doubt it would cause in your client's mind if they knew you had twenty years' experience but were still only billing at the rate of a junior. Sometimes you *have* to charge what the market thinks you are worth—for your clients to believe your worth. For example, if while I was making a booking to see a new dentist, I was told the consultation would be half-price, I would probably say thanks, go home, then phone back later and cancel!

I know a business analyst who was being contracted out at $65 per hour when the market was paying other people of his caliber (if they could find them) $150–$200 per hour. He wasn't especially motivated by money because he was content with what he had; his problem was that the work simply wasn't challenging enough. In his business, you don't get the best projects to work on if you're only charging $65 an hour.

He eventually solved his dilemma with just one short conversation. He told his client he wasn't happy with the rate they were paying him, "Come back with something more competitive." They did—twice his current rate and an infinitely more interesting project.

Price to save your clients the bother of doing it themselves

Another way to decide how much you should charge is to consider the amount it would cost your clients to provide the product or service themselves.

How much would it cost your customers to buy or lease the equipment, hire and train the staff, buy the raw materials, lease the building, package and transport the goods? How much time and inconvenience are you saving them, and what is that worth to them?

Price to meet objectives

Does your client have a problem he wants fixed, and he doesn't care how? Stop focusing on the task you normally fulfill and see it through your customer's eyes. What problem does he want solved and how much is that worth to him?

Listen for statements like these:

"I'd be happy to pay someone $600 a week to make my bed, shop, and cook for me and light the fire, so I can come to a warm house and hot meal at night."

"With the amount that dog has cost me in chewed shoes and furniture, I'd happily pay someone $1,000 to train my dog not to do that."

It might only take a professional dog trainer a couple of hours to fix the problem, and his return on effort would be much greater than if he were paid by the hour.

"I just want my place to look immaculate—I don't care what it costs."

There is a clear distinction to be made between the customer who wants to enjoy an "immaculate" home and the customer who pays a house cleaner to do three hours of cleaning and tidying. It may be to your advantage to quote to meet objectives rather than as performing a task.

Price according to risk

Stunt actors, oil-rig workers, and helicopter pilots do this. Call it "danger money," or put it down to the high price of insurance, or a shortened life or work-life expectancy. Lawyers also do this when they take a client on a pro-bono contractual basis, where they agree on a percentage of the client's compensation if they win the case.

In business ventures where you are willing to share some of the risk, you could suggest a pro-rata payment schedule, so that you receive a percentage on the results. Or, if you are in no position at all to take work on a speculation basis, ask for a retainer or a work-in-progress payment in advance.

Price according to providence

Another way of pricing is to simply ask for a donation—set no fees, and rely solely on providence!

I mention this pricing technique in a somewhat tongue-in-cheek manner, but there are plenty of bequests to charities that prove it works. Granted, it may be a little chancy, and it does require tremendous self-belief and/or trust in God, the universe, providence, or whatever you might like to call it. However, it has worked brilliantly for some churches, psychics, and other faith-based enterprises and operators.

Seek the highest bidder

Consulting firms often submit a tender for work, and Internet auctions have become de rigueur. Is it possible (and advantageous) in some way for you to put your work up for auction? In many ways, this is what happens to job seekers who sell themselves to the highest bidder.

Price to manage peaks and troughs

Many businesses have off-peak times—think about mobile-phone call plans and your local DVD rental store. Is there some way you can use varying price to generate more business during downtimes? Can you price to accommodate the flow of demand?

Perhaps you have a client with a non-urgent job who would be happy for you to call in when it suited you at a cheaper rate than normal. You could use this work to fill in gaps in your diary between other jobs.

Offer a free sample

Some companies give out free samples, or massively discounted products, in an attempt to "buy business"; they hope that the business they give away today will pay them back in multiples later, as customers discover how good their product or service is. It is an attempt to buy a future outcome.

If you are a multinational consumer-goods company selling something customers are likely to buy frequently in their lifetime (e.g., coffee or toothpaste), this strategy may pay off brilliantly; but if you are a small operator, it may be too risky, depending on your situation. If this strategy doesn't work for you immediately, how long or how often can you afford to play this high-stakes game?

Many life coaches give away free trial sessions. If you decide to give away a free trial, make sure it is going to be worth your while. Let's pretend your hourly rate is $100 and you have a track record of converting every second customer. By the time you have done one hour's paid work, you will have actually put in three hours (and that doesn't include expenses and other costs and overhead you are incurring during this time). Your hourly rate at this point is actually only $33.33.

Similarly, if your conversion rate is one in four, your hourly rate is actually only $20 at the end of the first paid hour. At the end of the second paid hour, you will have been paid $200 for six hours of work, so your hourly rate (with this client) at this point is just $33.33. At the end of the third paid hour, you will have earned $300 for seven hours, which is just $42.85. By the end of the fourth paid hour—$50—you're still a long way short of the $100 per hour you may be kidding yourself that you earn.

Another variation of the free trial happens in the advertising industry when clients ask agencies to "pitch" for their business. What this normally involves is thinking about the client's problem and solving it, all for nothing.

The reason agencies pitch is because it's a buyer's market and there are almost always two or three other companies prepared to "throw their hat in the ring," as the euphemism goes. One winner; two losers…and even then, sometimes the winner is the loser too because the business that they "won" never justified the costs incurred in the pitch.

An agency client of mine was asked to pitch for a trivial job that came with what I considered to be an "un-pitch-worthy" budget and no guarantees of any additional business. Here is my reply (though it never got sent!):

> Dear Mr. Spenda-notta-mucha,
>
> It was great to meet you at last. I appreciate Brian making the introduction and would very much like to help you, but after consideration of your brief, I regret to advise you that we do not provide speculative creative work without a guaranteed commitment to use our services.
>
> Unless the account is worth more than $200,000 dollars advertising expenditure per annum, we have a "no fee, no pitch" policy.
>
> In our experience, there is no such thing as a half-hearted pitch. To meet our own high standards, we would have to go through the entire process, incurring considerable costs, to create a product that has no resale value to anyone else.
>
> If you do wish to reconsider the terms of your brief, we would be happy to provide creative input for the mail campaign you discussed, at a cost of $35,000. Please refer to the enclosed quote for more details.
>
> Thank you again for your time.
>
> Kind regards,
>
> PS: Our copywriter wants to know if you'd like to come and spend the weekend tidying her office, for no money, and with no guarantee of any future business.

Suck it and see

Sometimes there is no predicting the emotional lengths the buyer will go to in order to get what they absolutely must have...super yachts, tulips, and love are just like that!

There's no market research in the world that can predict how many liters of a new beer will sell. Researchers can ask people what they *think* they will pay for a new beer in a can shaped like an armadillo, but in real life, the customer does not know himself until he's *actually* buying it.

Market research only tells us about past action, not future intent.

Until recently, the going rate for a lollipop was somewhere between ten and forty cents, but that was until they started advertising on television a triple-flavored pop-up version. If you had asked me five years ago, I never would have dreamed I would one day be spending $2.50 on a lollipop!

If you are offering something completely new to the market, or a variation of something old, you may simply have to do your homework, use your intuition, then launch it on the market and "suck it and see"!

Chapter 5

Determine Your Benchmarks

Last time I checked the latest range of business books and magazines on sale at my local bookstore, I left feeling deflated. It seems that unless you are building a business to sell or franchise, you're not doing it right.

But what if you don't want infrastructure? What if the only time you want to go global is with a surfboard under one arm and a set of skis under the other?

What if all you really want is to be your own boss, without having to worry about briefing staff?

What if you've spent the last twelve years of your life training to do eye surgery or sing opera or fly planes, and what you do can't be done by anyone other than you? Isn't there something honorable in that too?

The "grow your business and go global and franchise it" (GYBAGGAFI) tycoons will tell you that your business should be making money while you're not at work. They will frighten you with the thought that you're not doing it right—that your business might "just be a job." Oh, the shame there is in that! There are thousands of people who are employees and love having a job.

Next, the GYBAGGAFIs will scare you with the thought that unlike a business, you can't sell a job when you want to. Yet how many people have invested years in building the infrastructure of a business, and find themselves trapped until they find a buyer? At least with a job, you can just leave it!

And finally, the GYBAGGAFIs may try to make you feel insecure if something should happen and you're no longer able to work. But so long as you have insurance to cover you, are you any worse off than if you had been in a job? You may be unable to work and don't have a cash crop from a company to rely on, but have you noticed how the GYBAGGAFIs always assume you have the perfect moneymaking business and not the average-to-poor business that appears to be so prolific in real life? And if you're avoiding infrastructure because you know you'd be pathetic at managing it, which sort of business do you think you'd end up relying on?

Please understand that I'm not advocating the one-man band over the big-business model. If you love infrastructure and you want to build it, go for it. The point I'm getting to is this: If you are self-employed and all you have engineered for yourself is the monetary return equivalent to "just a job," you need to price (and insure) yourself so you can make financial provisions for when you're not able to work. If financial independence is your ultimate goal, you *must* price yourself right—and you haven't done that if you haven't factored in an allowance to take care of contingencies.

Financially, the bottom line is this:

1. If you are going to the trouble to go into business for yourself, you'll most likely want to earn at least as much as you could have earned if you were on a salary.

2. If you are going to invest your own capital in your business, you want to receive a return on your investment, or see some capital gain; otherwise, you're out of pocket. In financial terms, if your money is not being put to work for you, you are going backwards.

Larry is a marketing consultant who establishes an office, buys a car and computer equipment, and ends up investing $100,000 of his own money in start-up costs. If Larry had left that $100,000 in the bank, he would have earned interest. Let's say interest rates are 7 percent per annum; in that case, he would have received $7,000 income that year.

Therefore, if Larry is going to be businesslike about this, he needs to achieve $7,000 "profit" on top of what he pays himself, just to maintain parity with his comparative status as an employee.

Ideally he should return more than this, otherwise he could have taken the "safe" option of placing the money in the bank and earned a very easy $7,000 doing nothing. Furthermore, if Larry plans to sell his business to fund him in retirement, then his return on his investment needs to be a lot better than the interest rate the banks are offering—or why would a buyer (for strictly financial reasons) bother?

About pricing time

If this book was about pricing items for a café, we would at some point have to discuss portion size: how, for example, a large slice of carrot cake would cost more than a thin sliver, simply because the cost of ingredients was higher. However, this book is primarily about selling a service, so we are mainly talking about selling time. Though we may not have material ingredients as such, we do have portions of our time to sell.

We all have the same number of minutes in our day, so time is generally viewed as a limited resource. (For an alternative viewpoint, read The *80/20 Principle* by Richard Koch. In it he asserts that because we only make good use of 20 percent of our time, there is no shortage of it. Plus, if we doubled our time on the top 20 percent of activities, we could work a two-day week and achieve 60 percent more than now).

For the purposes of our discussion, let's assume we are operating at full capacity. If you are a bead-threader, you cannot make up wasted time if you are already consistently threading beads at your maximum rate per minute or hour.

Thus, if demand for your time is high, and you want to take advantage of this situation to make more money, you have a number of choices available to you:

- Work longer hours; never sleep!
- Get quicker at what you do
- Take on extra staff
- Subcontract the extra work and take a commission

- Increase your price so you do the same amount of work for more money…oh, happy days!

You need to think about the outcomes of each of these options, or discuss them with a business coach, as it is outside the scope of this book to advise you on your personal decisions related to how you want to do business and the choices you need to make about the way you put time and money to work for you.

What we need to do now is work out a revenue goal to find out if you are being realistic in your expectations: Will you be able to earn and charge enough to make it worth your while being in business? We need to dispel any illusions you may have about going into business and earning a million dollars a year, when there may not be enough hours in the day or customers in your reach.

About setting a profit revenue goal

When I did my coaching training, I learned about the SMART system for goal setting. I was taught that successful goals have these characteristics:

- **Specific.** Your wording of the goal needs to be carefully chosen and tightly defined, so it is clear what your objective is.
- **Measurable.** Your goal needs to be quantifiable, so it is clearly apparent if the goal objective has been met.
- **Achievable.** Your goal needs to be potentially within your actual capabilities.
- **Realistic.** Your goal must be enough of a stretch to be a challenge, but not so big that you could never attain it.
- **Time frame.** Your goal needs a deadline, so you hold yourself accountable and stop dreaming and start doing it.

Your goal might look like this:

"In six months' time, I will be working a thirty-hour week and earning $2,000 a week."

As we haven't yet discussed some of the factors that you need to consider when setting this goal, our goal right now is to gather the information you need before setting your goal.

Later on, you might like to come back and write your revenue goal here:

....

....

But first, we need to:

- Set a benchmark figure, which will be based on what you could have earned as an employee.
- Work out the fixed costs of operating your business.
- Figure out how many hours in which you can perform your work.
- Devise your break-even hourly rate—the amount you need to earn to cover costs and maintain parity with your employee earnings potential.
- Decide a profit revenue goal that will allow you to make the sort of money you want and still be what your customer is willing to pay.

Your benchmark figure

If you have work experience, a skill, or a qualification, and you are able to work, then I shall assume you could get someone to employ you on a salary and you have a very good idea of what your pre-tax annual salary would be. Include all commissions, bonuses, or profit share.

Write that figure here: $ .... [A]

[A] is your total annual salary before tax.

Now write down the value of any benefits you would receive from that employer. Include the annual value of things like health insurance, share options, superannuation, retirement schemes, gym, golf memberships, and so on. Include the value of a telephone and a company vehicle if they are for your personal use, but don't include them if they were 100 percent work-related expenses and you received no personal benefit from them.

Write that figure here: $ .... [B]

[B] is the total value of the benefits you receive per annum.

Forecast your fixed costs

Whether you're paying rent on premises or borrowing finances from the bank, you will be running up business overhead of some sort or another. In some cases, these expenses will be accumulating whether you're at work or asleep in your bed. To work out the yearly operating costs of being in business, write down all the expenses you will incur—this includes the costs of staff salaries, transportation, computers, telephones, postage and couriers, travel, office equipment, all insurance, rentals and leases, electricity, advertising, stationery, business advice, and finance costs (if you have to borrow money).

Take care with this, as I may not have mentioned all the outgoing expenses that apply to your unique situation. Do not include any variable costs[1]—those unique expenses or material costs that you invoice to each client on an individual basis. Remember, these are all the expenses you will incur for the privilege of being in business for yourself—none of which would bother you if you were still in that regular job!

Write that amount here: $ .... [C]

[C] is the sum total of your fixed costs per annum.

Now add [A] + [B] + [C] = [D] $ .... per year.

[D] is your benchmark figure. This is the amount you need to exceed if you are going to be better off (in purely economic terms) in your own business than if you had been drawing a salary working for someone else.

---

1. Note: Variable costs are those costs that increase according to the level of activity. This book concentrates on setting prices for tradespeople, or professional fees, so variable costs are not dealt with except to mention that variable costs should be charged to the client at the time of quoting and invoicing a job. It is up to the reader's discretion how much markup they may wish to put on these expenses, and no guidance on this aspect of pricing is given.

How many hours do you plan to work?

Let's work out how many hours in a year you can actually bill your customers. Okay…that's fifty-two weeks in a year multiplied by forty hours a week…no. *Stop!*

Your vacations were paid for when someone else kept you in a job, so you should maintain that standard when you work for yourself. Employers usually pay an annual salary for between forty-six and fifty weeks of the year, depending on the number of weeks' annual leave to which you are entitled.

Plus, just because there are twenty-four hours in a day and one hundred and sixty-eight hours in a week, you don't have to work all of them. How many of those do you want to work a day or a week?

The first step in calculating your hourly rate is to work out how many weeks you would expect to work in a year. Now deduct public holidays, sick days, vacation time, and study days. How many billable weeks do you have left?

Write that number here: ………………………… [E]

[E] is the number of weeks you can or will work per year.

Next, work out how many billable hours in a week you are able to work. As a guide, most of the big accounting and law firms expect their consultants to charge out between twenty-five and thirty hours.

Figure out the number of hours you can expect to work during the week without working yourself into an early grave. Be realistic. Allow time for traveling between jobs, and any other downtime you may have. If you are a one-man band and you are doing your own marketing and administration, you will need to deduct even more time to do these.

Write that number here: ………………………… [F]

[F] is the actual number of hours per week you can sell.

Multiply [E] x [F] = [G] ……………… billable hours per year.

[G] is the number of billable hours you have to sell per annum.

What's your break-even?

A break-even point is usually worked out on products—this is the price at which you would sell products only to recover costs while making zero profit. With a product, break-even points are often on a sliding scale, as production costs usually go down as the volume of products manufactured increases.

But this doesn't apply the same way when you only have your time and knowledge to sell. Instead, you have fixed overhead (operating costs) that accumulates regardless of whether you work one hour or one hundred hours a week. You can, however, work out a nominal break-even figure by totaling your fixed overhead costs for the year (those listed as figure [C]) and then dividing the number of hours you plan to work a year.

$$\frac{[C] \$ \ldots\ldots\ldots\ldots \text{ Fixed overhead}}{[G] \ldots\ldots\ldots\ldots \text{ Number of billable hours per annum}} = [H] \$ \ldots\ldots\ldots \text{ per hour}$$

[H] is the hourly amount you need to charge just to cover costs, or break even; at this point, you are making zero profit. [H] also assumes you are able to find enough business to fill all your available billable hours.

If you take on any work for less than [H], you now know the point at which it is actually costing you to go to work.

If a client pays you less per hour than [H], then you are effectively paying for the privilege to do work for that client!

Did the figure you arrived at for [H] surprise you?

A lot of people are surprised to find out the true costs of doing business. When you do the numbers, you can see why an employer may pay a staff member $50,000 a year but sets them revenue targets of $150,000. In this case, the $100,000 "profit" the employer has taken is not profit at all, but a return on the investment he has made in you. Some of this "profit" is eroded by expenses, and the rest is his "payback" for managing the risk and responsibility of being in business. Deduct a sum for the loss of business that may occur when you resign, factor in the costs of recruitment and training your replacement, and there may be very little left!

Keeping these figures in mind, we will move on to investigate whether your revenue goal (your break-even figure *plus* the margin you want to take in profit) is going to be realistic and achievable in the marketplace. What you learn in the next two chapters will help you identify your potential for profit, but whether or not you will be able to achieve your revenue goal is yet to be determined.

Chapter 6

Look Out for Number One

It is one thing to decide how much you want to earn, and quite another to achieve that in the marketplace. In this chapter, we will consider some of the variables—both opportunities and limitations—that may affect your business.

Imagine you are a river-raft operator and there are only a limited number of times you can go out a year. It costs you the same amount to go down the river with one person as it does with a full raft. Three people book, then show up with a friend—they argue that all four should go for the price of three, because your costs haven't increased. You argue that your price is based on an average price derived from all your fixed costs for all the trips you make, divided by the number of people during the season.

Let's say you change your mind and agree with your customers, and allow four to go for the price of three. What will you do when word has spread and only one person books, but three turn up?

What happens to your reputation when last week's party of six find out they were the only suckers to pay full price?

And what do you do when you fill your raft with non-paying passengers, and just as you're about to leave, a carload of customers appear, waving their credit cards at you?

This example serves to demonstrate why you need a pricing policy—one that will protect your profits and allow you to build your business for the long term.

Should you wholesale your time?

Consumers are usually rewarded for purchasing in bulk, and the purchasers of professional services (time) make no exception to the rule. But the trouble is that unlike cans of spaghetti, where the production costs go down as the quantity goes up, time does not work this way. Once you have used up today, you can't get it back again.

In general, the longer the project, the more likely it is that clients will demand a weekly or monthly rate, rather than an hourly rate. There is also a high probability that your competitors will meet this customer expectation, which puts pressure on you too to conform.

Among the benefits of wholesaling your time is the fact that you should be able to contract yourself for a set amount of time. The contract will provide you with some sense of security and allow you to plan ahead. But one of the pitfalls is that you may lose your presence in the marketplace. While you are "away," you may give your competitors time to encroach on your territory, or you may experience some downtime between jobs, if you have not arranged otherwise.

Under some circumstances, if the marketplace is short of people with your skills, you may find you can contract yourself to the client at a higher rate than usual, because the client sees value in knowing they have you "on tap." They may be willing to pay a premium to know they will never run out of what it is you have to offer.

Whether you decide to wholesale (discount) your time, or not is a judgment call on your part, but remember what I said earlier in chapter one: Don't be too quick to discount, and make sure that you don't use this as an excuse for not charging what you are worth.

Before you agree to wholesale your time, remember that time is a nonrenewable resource. Selling your time is what you do. Be mindful of what you agree to because while you are out selling your time to one customer, you cannot be dealing with another.

Whether or not you can afford to wholesale your time comes down to the economic goals you have set for yourself. Have you priced yourself in a way that allows you to sell your time at a bulk rate and still achieve your profit goal?

What units of time should you charge?

You may be wondering if you should charge by the minute, hour, day, week, or month. Obviously, I can't answer this for you, except to say: What does your client think? And which allows you to do the best job?

I phoned a man to fix the remote control on our garage door. Before he arrived, he told me his call-out rate was $25, with a flat fee of $15 to cover the first fifteen-minute consultation to assess the situation (payable even if the problem was unfixable). After that, he was eighty cents a minute. Clearly I was dealing with someone who was definite about his terms of business, and I knew exactly where I stood. True to form, he turned up exactly when he said he would, promptly fixed the door, and I was happy with a job well done, but I didn't make him coffee!

Think about what expectation you'll be creating in your client's mind if you tell her you charge by the minute. Will that frighten her off or reassure her?

Imagine being told that your architect charged $1,000 an hour. You know the architect's work will only be as good as the brief he gets, yet nevertheless, you know the clock is ticking. Which do you feel like doing: a) taking a leisurely stroll around your country estate, showing him the wild roses and bunny holes or b) waving a hand in the general direction, urging him to take a few photographs and telling him to get a move on?

If you know your clients will perceive your fees as high and you don't want to frighten business away before you've even got it, you may want to consider a one-off discounted fee (or free trial), so they can get to know you first and see the value in what you have to offer before they make a bigger commitment to you.

Would it be less threatening to the client if you apportioned your time into appointments? Remember, perception is reality; $100 for a fifteen-minute appointment *sounds* a lot more agreeable than $400 per hour.

Whatever you choose, think about performance issues—which of the units will enable you to deliver the best job and value to your client?

Should you charge by the job, hour, or percentage?

Imagine you have a client on the phone right now. "I saw your advertisement in the newspaper, and I wanted to know, how much do you charge?"

"What is it you want done?" you ask.

They may have expected to be told an hourly rate, but in one short sentence, you have turned the tables—now you have license to charge by the job.

I always prefer to charge by the project, because no two assignments are ever the same. Though I factor in an hourly rate when I do an estimate, I also take into account the size, value, and reputation of the client and the project in question. My fee takes into account timing, context, concept, and complexity. The more difficult the job, the more valued my unique skills are.

Obviously, not everyone is in the same situation, and when a customer calls, they expect a simple answer—either a price list for various services, a percentage, or an hourly rate. In some industries more than others, people like to "shop around," so it may be necessary for you to conform to the accepted way of doing things by providing them with an hourly rate; however, you still have the right to set parameters.

For example, if you are a plumber that unblocks drains, your hourly rate may be $60 per hour plus call-out expenses, under normal circumstances. However—and here are your parameters—your "drop everything and be there within sixty minutes" rate may be $120 per hour plus expenses. You may also have different rates depending on whether it's a day or night-time call, a residential or commercial call, a regular or new customer, and so on.

If you charge by the hour, the customer may gain a psychological sense of feeling in control of the buying process.

Mrs. Jones normally does her windows herself, but because she has hurt her back, she decides to call in a professional to do them for her. She looks in the telephone directory and calls three firms, choosing the window cleaner that is $5 an hour cheaper than the others. Mrs. Jones has the satisfaction of feeling like she may have got a good deal, though she has made some assumptions: 1) that all window cleaners deliver an equal standard of cleaning and 2) they all take the same amount of time to complete the job. She may not have stayed on the phone long enough to find out that one of the companies was run by an Olympic gold medalist window cleaner who cleans windows in a third of the time of his competitors.

If Mrs. Jones had gotten the three firms to come to her home and quote on the actual job, she might have found that the hourly rate was not the most cost-efficient way to choose.

In the area of professional services, many clients feel resistance to paying what they view as high hourly rates. For example, lawyers involved in setting up trusts or straightforward property conveyance usually provide the customer with a scale of charges. This provides the customer with a sense of surety about the budget, and nicely disguises what may in some cases be a very small amount of work for a relatively high hourly rate.

Some companies base their fees on a percentage of the value of the sale, to cover the labor costs and expenses involved in helping a client—recruitment, insurance, and real-estate agents are three professions that immediately come to mind. Like the price list, a percentage can sometimes disguise a very good hourly rate, but it can also mask the opposite. Often, agents are on subsistence retainers and commission, or commission only. Percentage pricing takes into account a more "swings of the roundabout" approach, which may not suit the unproductive, the inefficient, or the very unlucky. The fact so many clients buy from companies that operate on percentage pricing plans perpetuates the practice.

Years ago, I worked in a recruitment company where the client paid our fees on a percentage of the salary package the job applicant earned. There were some occasions when I interviewed the perfect person right way and made the placement. Easy money! But it was also true that there were other job vacancies I never filled after hours and hours of trying—and the clients in these cases were more than happy they weren't paying me by the hour. I know for a fact it all averaged out in the end.

Whether you charge by the hour, percentage, or by the job is determined in part by: 1) what the client expects, and 2) how much you can earn this way. Whatever you do, it is important to put your fees in context; for example, explain to the client that they pay three percent when you sell their house, but nothing if you don't.

How many customers can you manage?

A large retailer told me that even with the latest technology, his stores can only "process" one customer every six minutes. As he's already operating an extremely profitable brand, he can't see any point in spending more advertising money to drive more shoppers into his outlets until they're able to convert more shoppers into customers per hour.

Similarly, if you have a constraint on the volume of business you can handle—like the number of customers you can fit in a day—you may decide to work the numbers backwards by deciding how much you need to earn in a day or a week and dividing it by the number of customers you can fit in that time period.

What will the market bear?

Will customers be quite happy paying $350 for what you do, but stubbornly balk at paying you $500?

Let's pretend you are a forensic accountant with a large High Street accounting firm. You've been working in a large office with sea views; you've had a parking space, your own personal assistant, and four junior accountants working in your team. You always were incredibly good at what you do, and your clients know it. No one can get to the nitty-gritty of a fraudulent case quite like you do. But you hated the time-wasting meetings and the office politics, and now you've quit and are working from your home office, answering the phone yourself.

You're still the "big gun" you always were, but will your clients still see it that way? Now that you're working from home, should your charge-out rate still be $800 an hour, as it once was, or has something changed?

You may think that nothing has changed in terms of the quality of advice you dispense and the service you provide, but in reality, other subtle things have changed.

For instance, you are now out of the loop in terms of corporate knowledge. To keep up to date, you may find you need to attend more conferences and network more.

Now that you are at home, your client is no longer buying the intellectual property of the corporate brand they once were. What price do you think they may have put on that? They are also no longer buying the surety of the brand—the fact that in the past, people like you may have come and gone, yet the company has performed regardless. Will they have the peace of mind they are used to in terms of commitment to providing a backup if you are unavailable, and how much is that worth to them?

Your client may have liked the prestige and convenience of the High Street location and may have been willing to pay for that, and now that he knows your business doesn't carry the same overhead, how much less does he expect to pay? If you are going to do all the administration jobs yourself, then you may need to decrease your hourly rate, as you are now diluting your $800-an-hour thinking with $20-an-hour secretarial services.

To keep up with your workload, you may decide to employ a secretary and two bookkeepers. (Yours are more efficient because you don't waste their time in unnecessary meetings.) You may need to allow yourself an annual training and networking budget. You may also have to subscribe to professional organizations and journal subscriptions that when charged as an expense, work out to be more costly than when you were with the corporate firm because now you are the only person using these resources.

To decide whether you think you are still worth $800 an hour, you need to build a case in your defense, so you can give an explanation to your client. Perhaps you'll decide to lower your rates because your overhead is lower, or maybe you'll charge more because now you are going to get better qualified, specialize even further, and take on fewer clients, so you will become even more exclusive and valuable in what you do.

The value chain

Many service providers are "cogs in a wheel," or bit players in the production of a much larger product. Let's take as an example Mr. Evans, a property developer building a block of flats. Mr. Evans employs an architect, builder, electrician, plumber, painter, roofer, and so on. Each individual contractor has a business to run and wants to make their own margin, on to which Mr. Evans has to add his own profit before he sells the flats to a buyer. At each step of the way, each person wants to make a return on investment that justifies them using their time and financial resources.

This example shows why, if the service you provide is a part of a larger whole, you need to keep your eye firmly placed on what the end-use customer would consider a good deal. You may find there are times when you have to abandon some projects because you won't find any customers willing to pay the price you want to ask.

The value chain is where traditional mathematical sums start to go a bit funny.

If, in your customer's eyes, the value is less than you want to charge, the equation looks like this: 1+1+1=2. In other words, you have priced yourself out of a market.

Happily, sometimes the reverse is true, and your customer sees far greater value than the total costs of the ingredients—time, experience, costs, etc.—and is willing to pay it. In this case, the equation is 1+1+1=4 or even more!

Experiment with price points

Different price points will have different effects on revenue and profits. Your goal is to find the price point that yields the greatest revenue.

Some services are more elastic than offers—in particular, things that are subjective in value, like anything to do with beauty, art, and fashion. Personal taste is a fickle thing, so a sculptor may have more price elasticity to play with than a footpath concreter who simply lays concrete by the meter.

To illustrate a wide range of profit opportunities, I have therefore chosen a profession where there could be a market willing to pay a wide variance of prices, depending on whether they want their hair "cut" or "styled."

To work out your profit revenue at different price points, let's use a hairdresser who has just moved into the area. To attract new customers, she advertises ten-minute, after-school haircuts for children. She has available a maximum of twelve appointments per day, or sixty per week. Over a period of time, she tries three different price points—$9.95, $15.00, and $24.95—with the following results:

PRICE POINT	$4.95	$9.95	$19.95
HAIRCUTS PER WEEK	40	56	34
REVENUE	$198.00	$557.20	$678.30

In this instance, the largest revenue was generated at the highest price point, with the least amount of work. The $19.95 offer gives our hairdresser time to sit down and have a cup of tea or catch up on some phone calls between appointments.

Plus, if our hairdresser chooses the $19.95 price point, she will certainly be taking the healthier option in terms of reducing the RSI (repetitive stress injury) she might have gotten trying to cope with fifty-six out of sixty possible appointments in a week. She may however find that if she tested $14.95, she might have increased the number of appointments and made even more revenue, while at the same time "keeping the seat warm" for greater lengths of time. In the longer term, however, it may not be good for her reputation or for the customers' perception to see her getting only slightly more than 50 percent of her appointments filled at the $19.95 rate.

What attributes or assets can you leverage to increase perceived value?

Have you been voted most popular business in your neighborhood, or have you won any awards lately? Do you have a famous customer who has gone on record singing your praises? Are tickets to your shows sold out within minutes of first going on sale? Do crowds line up at the airport to greet you on your arrival home? Is there anything about you that gives you notoriety—anything at all you can use to your advantage?

What if your company is the garden-design team that wins a gold medal at the local flower show? How much more status do you think that confers, and what value will your customers see in you now? Would they be happy to pay ten percent more for your advice? Or twenty, fifty, one hundred, or five hundred percent more? This is time for you to test the market—time, maybe, to move in different circles, with clientele with higher expectations, bigger budgets, and fatter wallets…if that's what you really want to do.

If you do win an award or fame in some way and are not going to charge your clients extra for your "edge," you must at the very least let them know about your success and let them know they get "all this and more" when they hire you to do the job. This is one way you can show your clients your extra value.

What can you leverage for profit?

Can you copy the hamburger chains and ask your customers the equivalent of "do you want fries with that?"

Mary owned a beauty clinic and was receiving around sixty dollars an hour for her efforts. Most of her appointments took thirty or sixty minutes for full-body massages and facials, but the work was physically exhausting. After a treatment in her private room, she would take the client out to reception and smile to herself as clients would make impulse buys from her range of indulgent beauty products and perfume. She noted how ironic it was that in the same time it took her clients to sign a docket for the sixty-dollar facial, they could sign one for one hundred and sixty, buying a hundred dollars or more worth of products with a 40 percent profit margin. And the best thing about that profit was that it took her no extra physical effort.

Work out the cost of your conversions

If all else is equal, the salesperson who converts ten out of ten calls for quotes into orders can actually *afford* to do the job cheaper than the salesperson who only converts three quotes into sales.

How much time can you afford to spend between jobs? Your fixed costs won't go down just because you are not working. At some point you will need to decide how long you are prepared to hold out for a higher price.

The more time and money you spend trying to win the business, the more costs you'll have to factor into the price of the business you do get.

Promotional pricing and repeat business

Advertising or spending time and money to find new customers is a lot more expensive than keeping an existing customer. Along the same lines as the saying "A bird in the hand is worth two in the bush," pricing your goods or service to guarantee repeat business makes good commercial sense, which is why cafés and barbers are so keen to give you their "Buy five, get one free" vouchers. Over the course of a year or longer, this can add up to a substantial amount of business.

While a promotional (discounted) price might entice new customers and temporarily spike sales, there is a downside if your customers refuse to buy from you at the normal price. You will need to make sure you factor the overall cost of all that discounting into the equation, or it will jeopardize your profit goals.

Always be vigilant that your promotional price is helping you increase sales enquiries that are converting into sales. If all it is doing is generating enquiries that are taking you time to respond to, it is actually keeping you busy and costing you money.

Check also that you (as the vendor) don't get addicted—the continuous use of promotional pricing may be a psychological foil for never charging what you are really worth.

The truth about discounting

Discounting to increase sales volume may work well for products because it's a mathematical reality that you would generate $20 more profit if you sold one hundred items at $1 profit rather than forty items at $2 profit. However, if you only have your personal time to sell, one hundred hours is one hundred hours, no matter which way you try to cut it.

There are several problems associated with discounting, but one of the biggest is that it focuses everything on price—as if that were all that matters. If your only competitive advantage is price, you are in trouble, because price can always be matched.

Discounting starts price wars. The company that usually wins is the one with the biggest balance sheet—the one who can afford to hold out the longest.

Discounting can also affect the customer perception of your service. Consider what would happen if all your competitors met your discounted price—do you think your customer is going to accept any less quality?

Your client won't discount his expectations, yet if your discount affects your margins and you are forced to compromise the quality of what you sell, you may lose repeat business, get a bad reputation, or end up spending time fixing complaints, and so on.

Ten easy ways to discount and lose money, FAST!

1. Don't explain your terms of business before starting work
2. Wholesale your time
3. Don't have a price policy in place, or do not stick to it if you do
4. Make spur-of-the-moment guesses instead of pricing things properly
5. Work for friends at "mates' rates"
6. Don't have any convincing sales reasons for not discounting
7. Start work on the job before the client has confirmed terms and conditions
8. Underestimate the amount of work involved
9. Overestimate your ability (to get things done quickly)
10. Forget to write things down; keep poor records

One of the basic rules of negotiating is that if you are going to offer a discount, you offer a different product or service. Challenge your customer's proposition for a discount with: "If you want a better price, give me a better order." Perhaps negotiate different terms, or a shorter guarantee, or longer lead times.

Earlier, I wrote about another form of discounting in the form of offering a free trial. If your clients are asking for discounts, you need to know how to negotiate: how to counter their demands (which are stacked in their favor) with an offer that is stacked in your favor. The ultimate outcome is that you both do business and both feel good about it; this is called a win-win situation. Learning how to become a good negotiator is a topic that entire books are written about, so I suggest you read one of those. In the meantime, I'll give you twenty useful tips to help increase your negotiating power because the better your negotiating skills, the easier you'll find it to represent your best interests when dealing with a client who wants you to discount.

Twenty tips for negotiating success

1. See it from your client's point of view

   Step inside the shoes of your client for a moment and think about her fears. Does she know she can trust you? How does she know you will deliver good value for her money? Will you make her look stupid for buying from you?

   Make sure you understand exactly what she wants and what she expects to receive from you. Invest time in educating your client about how you operate.

2. Plan in advance

   Make a list of what you have to negotiate. What will your customer insist on, and how flexible are you prepared to be? Can you offer delivery before or after Christmas? Can you offer a six-month or twelve-month warranty…or would he prefer a monthly contract? Look for solutions, not problems, and keep thinking creatively. Ask him: "Would you prefer this, or what about that?"

3. Make sure you are dealing with the person who has the authority to say yes

   If possible, find out how much discretion the person you are dealing with has, and who else is involved in the purchase decision. How are you going to sell them anything if the person you are dealing with has no authority to buy?

4. Don't mention your price too soon

   Get your client thinking about what you are going to give her before you start her thinking about what she's going to have to pay you. Build her confidence and reassure her that the quality will be just what she wants, and make sure it is, or let the client go.

5. Test them

    Tell your customer what others are paying and how happy they are with your products and service. Then test your customer by asking: What were you expecting to pay? How much are you paying now? What are your usual rates? Is there a company policy on this? What did you pay the last person?

6. Sandwich your price between benefits

    When you tell your customer how much it costs, tell her the significance of all that she is getting and how it will benefit her.

7. Compare the product quality to the price difference

    If you are making a comparison between your own products and services and those of a competitor, tell your client all the extra-special effort or components you put in. Remember: it's not the price that's important; it's what the product does for her.

    If your product or service is higher-priced, don't be afraid of the competition. Be proud. You are better. Likewise, if you are cheaper—be proud. Tell your customer how and why you made it cheaper, and what the benefit of this is.

8. Show the penalties of not buying

    What might she miss out on if she doesn't buy now?

    Show the savings—gross them up or show them as extra profits if they are reselling your goods.

9. Explain the cost (or the savings) in a way that is meaningful to your customer

    Reduce the expense into smaller units *(e.g.,* $520 extra is just $10 a week, which is $1.42 per day, or a few cents per hour). If it's a large purchase, you can work it over the lifetime of the product.

Compare the value of your deal: What else could they spend their money on? Put it in their language. For example, the cost is less than one meal out. You only need to make one sale to cover it.

10. Create some urgency

    Motivate your customer to act quickly. Make them an offer. For example: buy one and get one free; free gift with purchase; extended five-year warranty.

    Focus their attention on deciding now, not later. Give them a deadline on the offer. "Offer valid until (date)." You don't want them to procrastinate any longer.

11. Sell the discounts

    Some people just love a bargain; in fact, they *must* get a bargain. Keep their attention on the $300 they will save, not the $6,000 they will be spending.

    Most people relate to cash better than a percentage—three hundred dollars is far more meaningful that five percent!

    Use price points to ease your customer through the psychological pain barrier—$9,999 looks a lot less than $10,000.

12. Ask them to make a counter-proposal

    Get your client to bring their ideas into the open. See how far apart you are. Do a reality check: go back to the original brief and specifications and check whether they are all mandatory. Since starting the purchase process, it is possible that your client may have altered his expectations.

13. Price bargain

    "If you want a better price, give me a better order." Search for joint benefits and win-win solutions. For example, you could lower the price if the customer does part of the job or pays you cash in advance.

It is in the nature of your customer to want all the discounts you give him. It is not his job to tell you, nor is he ever likely to tell you, that you're too cheap.

14. Don't drop your price before it is necessary

   Don't become too committed to your lowest price early on. Try to hold your price by giving them more value—an extended warranty, a free upgrade. You can always come down later, but it's difficult to go up.

   If you give them something, make sure they appreciate you more for it.

15. Leave yourself room to negotiate

   Allow for contingencies. Give yourself the right to add or alter the quotation later if the client changes their mind in some way, or you have unexpected difficulties sourcing raw materials and the like. Put your conditions in writing to avoid any misunderstandings later.

16. Keep your objectivity

   If the negotiation is overheating, take time to think things over. Be wary of high-pressure tactics. Give yourself an "out clause," like referring to a higher authority (your boss, your lawyer, your mother), so you have time to take the deal away and review it.

17. Practice saying no, maybe…okay then

   Negotiating is a two-way process. You can dance to and fro for ages, but always keep in mind the bottom line and be prepared to walk away. Use your intuition, and don't allow yourself to be bullied. You can always say no, but do it in such a way that you leave the door open, so you can change your mind and say yes later if you think you might want to.

18. Listen for the ticking clock

   One property investor buys a house every year in the week prior to Christmas. She gets a great deal, and the owner goes on vacation able to make plans for the future. Make sure you know your client's timeline and use this information to your advantage.

Also make sure your contract has specific time frames; *e.g.,* "This quote is valid until (date)." You want to retain control over your time—this is your life.

19. Assume nothing

   When you present the quote, some clients go quiet for a while before responding. Others react as if they've never seen or heard anything so preposterous. You can never know for sure what they're thinking. Ask as many questions as you can, to try to find out and be confident that you know why you charge what you do, what's negotiable and what's not.

20. Keep an open mind

   If you find the market situation or your competitors' tactics have changed, you may need to recheck your facts. Don't believe that you offer the right price for all time—do some research and intervene, if necessary.

Chapter 7

Decide Your Value

You know what price you need to get to stay in business, but now it is time to get down to brass tacks: how much is your customer willing to pay? Will there be enough volume of business for you to earn the revenue goal you set yourself? How are you going to find out?

The first thing to remember is that this is not rocket science. To find out how much it costs to fly to London in May, wouldn't you look in the paper, phone or call into a travel centre, ask your neighbor who knows absolutely *everything*, and maybe do an online search on the Internet?

To find out the "going rate," you should do the same. Act like a customer and do your homework on your business colleagues or competition. If it's not appropriate for you to be seen "scouting" the territory, recruit a friend or family member or pay someone.

Make sure you stay focused on what you need to know by drafting a telephone script or list of questions *before* you or your assistant investigate any of these helpful sources.

In addition, you should glean what information you can from authorities like your accountant, lawyer, recruitment consultant, business sales broker, and any other professional advisers who know the workings of a business like yours. Read the industry journals, join any relevant professional associations, and network at seminars or meetings.

Ask your customers

Depending on your situation, you could hire a market-research company or arrange to do some yourself. Invite a sample of your customers to a focus group. Make attendance at the event attractive in some way—perhaps offer a gift or discount off a future purchase—but be careful not to bias the results. Make sure you get a range of customers from best to worst; you'll learn as much from your detractors as from your fans.

Ask your customers what they think about your product, service, or price. Find out what they like and dislike about you and your competition. What could you do to make your customers happier? Perhaps you've got some samples or new ideas you'd like to run past them. Would they pay you more for something? What would they like that you don't currently offer? What could you change, and what are you best known for? You want to find out what is and isn't working for you—which aspects of your business your customers do and don't want to see changed.

Be aware that customers can only tell the researchers what they've done in the past and what they think they'll do in the future—they do not know themselves how much they might actually pay for a piña-colada-flavored lollipop that sings "Happy Christmas" until they are actually buying it!

If you don't have any previous knowledge of market research, I suggest you read a more specialized book or enlist the help of an expert to make sure you get the most accurate information for all your efforts. The planning you put in at this early stage could be a major part of your success later on.

Get a second opinion

Try selling the price of your service (and/or products) to someone like your accountant—he or she may point out some variables you have forgotten to factor in.

Get a second opinion from someone who has never bought from you, a complete stranger.

Now get a third, fourth, and fifth opinion

Next, present your product and price to a random selection of people whom you can trust to give your project serious thought. When you try answering their questions, listen to hear how strong or weak your price arguments sound when you say them out loud.

Perhaps try your spouse, mother, or neighborhood tarot card reader. Keep your ears open; don't judge what they have to say. Make sure you don't set out to prove yourself right; keep an open mind to new viewpoints.

Test the market

Another option is to run a test trial. Find a small pocket of customers you can try your new ideas on—just be very careful that you do it in such a way that you don't damage your credibility for the future.

If possible, try selling your product at higher and lower price points to see if it makes a difference. You never know—you may discover you can put your price up *and* attract more customers, all in one go!

Decide what suits you

One of my absolute rules for doing business has always been that I must like the person I'm dealing with. I also have a minimum profit objective I set out to achieve, along with some other quirky "rules" that are personal to my work ethic and situation.

Those rules suit me, but may not be appropriate for you, so you need to decide what suits you. What rules do you want to work by? What are you going to do, to what standard, at what price, and for whom?

Your pricing policy will be different from mine. How much you earn is only one piece of your overall financial picture, so you have to juggle your own costs and profit intentions. Only you can know what will motivate you and make you feel proud and happy. Equally, only you can decide what is non-negotiable.

If you have diligently worked through the previous chapters, you'll know what your break-even point is. You'll know your floor price (the lowest price you can accept and still remain in business) and you'll know the price ceiling (maximum) your customers will pay.

So decide! How much is your price going to be, and who will you do business with, and what will you or won't you do for the money? Ultimately, what you charge will represent both your value and your values.

Loosen up; get creative—who says business always has to be so serious? See if some of these examples trigger something in you:

"I won't get out of bed in the morning for less than $1,000."

"I won't invest my own capital in the business without getting a better return than 33 percent."

"I only start work once I have 50 percent of the money up front."

"I won't be beaten on price—whatever the other guy charges, I'll do it for less. I'm going to be the market leader at all costs." (Good luck to you with that one!)

"If they want it in a hurry, they can pay 20 percent extra."

"I will not take on any job unless I earn $25 an hour, and I get a greater than 20 percent markup on materials."

"I never pitch. I don't care if someone else will do it for nothing. I don't pitch."

"I will work for $280 an hour and I will only work for people I like."

"My hourly rate is $12.50 an hour with a minimum of two hours, plus travel expenses at $1.00 per kilometer."

"If they want me, they can have me, but only on an annual contract of $250,000 plus travel expenses."

Take a moment now to write down the conditions that are important to you:

....

....

....

....

CHAPTER 8

Present Your Price with Confidence

If you completed the last chapter, well done: congratulations! You've done the hard yards in terms of decision-making. Now it is time to improve your presentation skills to make sure you increase your chances of winning the business you want.

If you were a product in a shop, you might display a price tag, which would simplify things a great deal, but you're not. In your situation, you're going to have to ask for the money you want by presenting a written quote.

As a minimum, your quote should include your business name, address, and contact details; your customer's name and address; a detailed description of the work you are providing a quote for, along with a price (including the costs of any materials and other expenses); and an expiration date, so your client doesn't try to take you up on your offer after you've experienced increased expenses of your own.

Depending on the business you are in, you should know the best way to present your quote, be it in person or via e-mail, letter, fax, or telephone. Generally, the more expensive, complex, or technical the job, the more personal you should make your presentation. For example, if I were presenting a quote for a new client on a big new campaign, I would show up in person and explain the quote; but if it were a tiny add-on job of low value with an existing client, I would send an e-mail message and follow up with a telephone call.

You need to work out which presentation nets the best result. Is there more you can do at the time you present your price to increase your chances of making the sale?

For example, have you (concisely) emphasized the benefits of them doing business with you? Perhaps you should send a small gift along with your proposal—something small that won't be construed as a bribe, perhaps a pen bearing your logo or a small bunch of flowers.

Is your quote too long-winded or too brief? Have you asked your customers what they think?

It's not what you say, but how you say it

Have you ever noticed how easy it is to ignore people that approach you with the "why can't I have that?" tone in their voice, whereas you feel inclined to enter into discussion with people who greet you as an equal and are upfront and direct in their requests?

I noticed this recently when my seven-year-old came demanding money. There was something in the pitch and tone of his voice that annoyed me: "Muuum, can I have two dollars?"

My reply was, *"No!* And besides, you didn't even say please."

A day later, he asked me, "Mum, how can I earn two dollars?"

On both occasions he had the same objective, but with two different results.

Practice, practice, practice

Your aim is to become as professional at asking for the money as you are at doing the business, and the best way to do this is to set aside some time to practice.

Write a script and rehearse it until the words that come out of your mouth are the "perfect" expression of how you want to sound.

Practice talking to yourself in the car or in front of the mirror so you are comfortable talking about what you do and how much it costs.

Communication is more than what you say: Is your body language confident? Are you looking your customer in the eye, or are you fiddling with your collar, looking untrustworthy? Go and read a book on body language if you have any doubts.

Find a work colleague, a friend, or a professional sales trainer or coach to practice with. Role-play some negotiating sessions, videotape them, and play them back later—see for yourself how you could improve how you look and sound.

Check if you have some past customers or trusted fans you can ask for feedback—maybe take one or two out for lunch and ask for their candid opinion—but only do this if you're sure you won't jeopardize your credibility for the future. Set aside some customers to practice on, and if you feel comfortable enough, ask them for their feedback. Ask for comments from customers that chose not to buy from you; learn wherever you can how to improve your conversion rate (*i.e.,* the number of enquiries you convert to actual sales).

In addition, if affirmations work for you, do them. Go and stand in front of a mirror, or talk into a tape recorder, and repeat after me: "Hello, my name is _____ and I'm a recovering under-charger. Every day, in every way, I'm getting dearer and dearer…"

Okay, so I'm jesting about "getting dearer and dearer." My message is not necessarily about charging more and more, to the point that you start ripping people off. You do, however, need to reprogram yourself to get to the stage where you believe in yourself enough to charge what you are worth.

For example, "I do an excellent job and I am worth $150 per hour." Or: "I am good at what I do and I'm worth what I charge." "I gratefully accept all tips customers want to give me."

Have your answer ready

If the client asks you how you set your prices, it will make your life a lot easier to have a well-prepared answer ready.

To illustrate, I have taken an excerpt from a short story I wrote about a management consultant, Trent, and his client, Cecil. It may be exaggerated for effect, but it serves to make the point:

"So how do you set your fees?" asked Cecil.

"I price my time on what I think it's worth," Trent replied.

"And how do you work that out?"

Trent shifted in his chair, his eyes lighting up. "Now *there's* a leading question! How do you price your precious product?"

Cecil smiled and grunted. "Hmmph, I could tell you that, but then I'd have to shoot you!"

Trent laughed. "Precisely! It works the same way for us—pricing is an integral part of our success."

Cecil said nothing, so Trent continued. "Actually, we do have a formula back at the office. I record all the time and costs involved in a project, then I load them into a spreadsheet and multiply everything by eighteen and divide the result by six and three-quarters and add on $4,500."

Trent acknowledged Cecil's laugh, then adopted a more serious tone. "Joking aside, I will give you some explanation—it may give you an insight into our values, and the way we work. Basically, I run my hours like a business, recognizing that like any business, I have a limited supply of raw materials and variable demand. And like any product, I have a limited lifespan—a use-by date, in other words."

Trent cleared his throat, and Cecil smiled and nodded for him to continue.

"I need to factor in downtime and costs of repairs and maintenance (you might call yours vacations). I also allow a sum for upgrades—I network with others, so I can access their abilities when I need them. I also feel entitled to make a return on the investment made in my education, plus I allow for ongoing updates. Then I add in a bit for contingencies and deduct a bit for noncompliance (you probably call yours bad days). Do you see how I'm thinking?"

Cecil smiled and nodded.

"You see, it is my 'edge' that I value the most, and price accordingly. My edge is my magic, my insight, what you get when you retain me. And I have to price that at a premium to ensure that I respect and value my contribution; it's easy to devalue that and overexpose and overcommit myself in the marketplace. Exclusivity comes at a price, and I have to self-manage to ensure that I take enough time out to fuel my creativity and retain my objectivity, which is what clients are paying for."

Trent paused and fixed Cecil in his gaze.

"The truth is, I think everyone should see their time as a tradable commodity, regardless of whether they've got a boss or are self-employed. At the end of the day, it comes down to self-worth and what price you want to put on your life. Pricing is a discipline—but I think from our discussion, you know that already."

Cecil nodded. "When you put it like that, it makes me start to worry I can't afford you."

"Can't afford to take me on…or can't afford not to?"

The words "opportunity cost" hung invisibly in the air above them.

Finally Trent broke the silence. "We both know you could pay a lot less but only get a fraction of the results."

"Good point," said Cecil, "but the bit about results—how do I know that for sure?"

"You'll know," Trent said firmly. "You always know when you've bought quality."

Be prepared for the difficult moments

Always remember that life is about decisions, not delusions.
—Pernelle, thirteen years old

Life isn't always about plain sailing, which is why I've included this section on what to do when the discussions between you and your client don't quite go according to plan.

Let's start with the client whom you're getting along well with, right up until you mention the price, when they try to make a run for the door.

Some clients won't raise the money issue until you do. Until you get to the point of discussing money with them, they will feel perfectly comfortable, but anything after that frightens them. In these situations, it is best that you take the initiative and bring any issues to do with price and payment options out into the open.

If you are doing the selling, the onus is on you to guide the customer through the buying process. Take the issue of price head-on.

Be matter-of-fact about your terms of business—perhaps they can open an account with you, or maybe you insist they pay one third up front, or cash on delivery, or payment on the twentieth of the following month.

Perhaps you need to help them arrange financing so they can afford you. If it's worth your while, go the extra mile and do that for them.

For example, I recently discovered that the "going rate" for orthodontists is $5,750 per mouthful of braces, regardless of the age of the child, the number of sets of braces they're going to need as they grow, or the degree of teeth straightening required. Instead of giving the traumatized parent a cup of tea and a biscuit to get over the shock, they receive a monthly repayment plan that operates even though some of the appointment regimes are only biannual. There's no need to split hairs over time spent per child and number of consultations, just "one price fits all" and an easy repayment plan to accompany it.

But what if they object to your price?

First up, you always have the option of saying "Too bad" and moving on to the next client. I doubt you'll have read about this technique in very many gung-ho "How to sell potatoes to the Irish" types of sales books—but it is an option nonetheless. It may even be a very good option if you don't mind long odds. You may have seen this in action at a shop where you've looked at the price of an item and thought, "They must be joking." The fact remains that every now and again, someone does pay the asking price, making it all worthwhile (to them).

But let's assume you're happy with the amount you're asking for; you just need to address your client's query. Usually, any request for more information is a buying signal, so don't get defensive, but do check that the price is the real reason they are not quite ready to buy; there's no point in going any further until this is sorted out.

Ask them: "Is this exactly what you want if only the price were right?"

Deal with other issues if they come up, because often the reason the client won't buy is that they don't see the value. It is possible the client may have gone "cold" on the purchase altogether; otherwise, if price really is a sticking point (they would buy from you if only you were a little less expensive), here are a few replies you may like to try:

"What price did you have in mind?" Followed by: "Where did you get that figure from?" Find out if they know something you don't. Keep asking them questions to estimate how close to your asking price they may be willing to come, then decide if you want to do a deal on those terms.

Or you can stand by your price and reaffirm the value. Let them know you are not negotiable; they must pay your price if they want to get you; they cannot get you any other way.

"I know we're not the cheapest, and we never aim to be." Now tell them all the reasons you are better qualified, and how you provide more value.

"We don't cut corners; all our staff are fully qualified; we've invested in better equipment; we've been in business fifteen years and plan to still be in business in ten years time; we always keep our word; we guarantee our work; no one does it better…"

"As the saying goes, you can have fast, cheap, and good, but not all at once. You can only have two of those things; which two do you want to pick?'

"The other company must be using some very cheap materials…where do you think they might be skimping?" Or: "You know what they say: you get what you pay for."

"I see. Well, of course, it may be possible to reduce our price. Let me think if I can find some ways to cut corners."

"Which part of the job don't you want me to think about?" Or, "What can we eliminate?"

Perhaps you're not compatible

Sometimes the fact is that you simply cannot provide whatever it is that will make the purchaser happy. It just may not be viable for you, or you may simply choose not to meet their demands.

For several years I worked with a large client who used a prestigious advertising agency for the big-budget televised promotions, and me for the smaller direct-marketing projects. The client approached me soon after I won a bunch of awards and was happy to work with me, knowing that I worked from a home office but that they would receive high-quality creative work, on time and on budget. To keep the client satisfied, I always negotiated a time frame that was reasonably prompt, but also realistic for me. This worked well for a couple of years, until the marketplace and the client's expectations changed and it all became increasingly stressful. I could not—nor did I ever want to—deliver as quickly as my client had come to expect from his other agency.

I saw the light (and the lunacy) the day I was in his office and he wanted to show me the artwork for a new television campaign that was about to air. On his desk he had two small stacks of documents; he lifted the top page of each pile, couldn't find what he was looking for, and picked up the phone and asked someone at the other end to send him another copy. Putting the phone down, he smiled at me and said, "It doesn't matter. XYZ Advertising Agency has a ten-minute courier. I might as well use it."

Talk about teaching your client how to treat you! I guessed it would have taken thirty seconds to have found the artwork on his desk, but I returned his smile, hiding my annoyance, knowing that some poor sucker would now have to stop what they were doing, reprint the items, and deliver them. Sure enough, by the time the ten-minute courier arrived (fifty-five minutes later), my car had a parking ticket for exceeding time.

I could see this new ten-minute courier was a threat to my operation in terms of my client's changing service expectations. Being newly pregnant at the time and having no intention (not at any price!) of trying to equal this service, I soon resigned the client.

Facing rejection

Rejection is a fact of life. Sometimes you get to be the rejecter; other times you are the rejectee; but in either case, life goes on.

What did you do last time you bought a car? Did you buy the first one you looked at, or did you look around a bit, read the papers, and get your "eye" in on price? Would you get more than one quote if you were printing a large run of brochures to drop in letter boxes in your area?

Some companies have a policy of asking for three quotes; that means two quotes are going to be unsuccessful—boo hoo—but it's not personal.

What to do if the client gets personal

I once had a client that would make personal comments whenever I wore anything new or upgraded my car or went on holiday. He seemed to be implying he had paid for them. I remember him grabbing my hand, peering at the cubic zircon (gorgeous, but not valuable) ring my mother had bought me. "Are those diamonds?" he asked, breathing heavily on my hand. "They are, aren't they?" he quizzed. "Phewweee! We must be paying you too much."

I felt resentful because it wasn't as though he was my only client. How much I earned was none of his business. Besides, I had a hardworking husband, and I had by my previous efforts accumulated some wealth of my own. (Observe here my beliefs in money and hard work!)

As I was still building my reputation and work portfolio, I wanted his business— but what a struggle! It was his money mentality I was up against.

It often seemed so unfair, I would be given punishing deadlines, and then it was implied I was expensive. My mind would rebel with an internal tape that went like this: How dare he? It's all right for him with his company car, expense accounts, health insurance, and international conference junkets—he should try earning a living just as I am in the "real world."

My problem was that I really needed to read a book like this one. I was insecure, and, as personal as the client was getting, I was "buying into it"—not that I realized I was doing that at the time.

Now that I know better, I realize that this mind game only works if you allow yourself to be susceptible to it. These days, I would probably respond with humor. "Yes, it is a beautiful ring, but I do find my finger gets tired under the weight of it!"

Sometimes clients working for corporations would themselves like to break free and do their own thing, and they envy what you are doing. Other times they may genuinely be ignorant of the true effort and costs involved in running a viable enterprise of your own—after all, they haven't been there and done that themselves. Sometimes they fear you are earning more than they are—just the same way they fear their colleague in the office next door might be. They might think it's too easy for you, that you're not struggling enough; but whatever your client thinks, they have a perfect right to think that way. Let them express their money mentality so you can tune into it—or tune it out!

How to respond if they ask for a discount

First, ask yourself: What effect will discounting have on your reputation? What effect will it have on profitability…and what about future sales? Could it rob future sales as they stockpile at cheap rates, or will it set a precedent—a new low for you and others in your industry?

Why are they asking for a discount? If you are being asked for a discount on their normal order, why would you do that? Save your discount for when they do something special for you—like, for example, giving you exclusivity.

If you are tempted to think of your discount as an incentive, is discounting the best incentive you could offer? What about offering a better product, or an extended warranty, or something else?

You may believe that you have to discount, because everyone in your industry does it. Maybe the fact that you don't discount (you don't have to) is your point of difference, and pride. Don't forget the fact that many price wars are started by customers.

Perhaps you decide to offer a discount to outdo a competitor, or as a way to match a competitor by discounting to "save face," but always be aware of the strategic implications of discounting, and be certain you can defend your actions.

Alternatively, if you know your client is likely to ask for a discount, build a margin into your first price so that when you play the "discounting game," the price still meets your own profit criteria.

Prepare yourself now, by writing down three reasons you could give to a client as to why you don't discount. Now write down conditions under which you would, making sure you phrase these concessions in a way that makes it clear you're giving them a special reward. Make them appreciate it.

What if your client tries to undermine your competence?

Another way clients get more for their money than you bargained on giving is to query your efficiency or competence.

I don't know anyone who wants to be seen to be inefficient or incompetent, so it's only natural to react by wanting to cover it up. This can become a problem for you when the allegation is incorrect or unfair and you react unconsciously, trying to change that person's opinion of you to restore your good name.

Even nice clients do this, and sometimes I'm sure they aren't even aware they are doing it. The reason they do it is because they get such good results (the pay-off) that they may not even want to change their behavior—and why would they, when it works so wonderfully for them?

Watch out for the client who tells you Henry can do it in half the time. Simply smile and tell them how long it will take you and how much it will cost. Alternatively suggest that they take it to Henry, as it could work out to be more cost-effective for both of you that way.

Beware also the client who asks: "Can you fix this up for me? It's only a simple little job."

I have an acutely developed ear for the words "it'll just" or "it's only." I trust my instincts and know that it means one of the following:

- The client wants to minimize the job, hoping it'll keep the price low or get the job done quickly, or both.

- The client is ignorant and either doesn't know the value of the job, or doesn't think he needs to know the details. Either way, he needs to be educated.

- The client thinks she is the expert and is trying to let you know that she knows more than you do.

- The client presumes to know your time frame and wants to run your business for you.

How to deal with friends who want to work with you

According to marketing gurus Al Ries and Jack Trout, "the number one reason for using a company is familiarity," yet conventional business wisdom tells us not to do business with friends. In my experience, this situation is difficult to avoid, particularly as I work in a niche industry and have worked intensely with clients year after year. Over time, I do end up befriending some of my clients, especially as one of my rules of doing business is that I only work with people I like.

There is good reason for doctors to not treat family members, and there are times when it is inadvisable for you to work for a friend, like when you are emotionally involved and your judgment might be impaired in some way. But if those conditions do not apply and a friend wants to hire you, then why not go ahead and treat them like anyone else? They are probably asking you because they respect and trust you, and you are the best person for the job.

The most important thing is that you explain and document your terms of business, and take extra care to communicate honestly if there is any likelihood that one of you could be operating under false assumptions.

A while ago, my commercial artist friend Neil did a job for his friend, a dentist. Neil gave his friend a $200 discount, but when he went to get a root canal done, he received only a $50 discount.

This time, when Neil told me he was going back to get his teeth checked, I couldn't resist asking him how much discount he was going to get this time.

"Same as last time: 5 percent," he replied.

Without thinking, I said, "That's a bit tight." A moment later, I realized what I had said and felt embarrassed. It just goes to show how old ways of thinking die hard—I could write a book on this subject and still not have ironed out the wrinkles in my way of thinking!

Feeling chagrined, I thought: Who am I to judge 5 percent as mean? Isn't 5 percent a lot when you consider the dentist has a busy practice and could have seen a full-price customer instead of a friend?

My only excuse was that I was empathizing with my friend, and I was ignorant of all the facts.

"How do you know he gave you 5 percent?" I asked.

"He added up the consultation fees and costs, then deducted 5 percent from the total," Neil replied.

Thinking about his original $200 discount, I was curious. "So when you invoiced him, did you itemize your discount the same way?"

"No," said Neil.

"So he had no way of knowing you'd given him $200 of work for free?"

"No," said Neil.

Learning this, I felt even worse about judging the dentist badly! Neil and I then discussed the consequences of ad hoc and unequal gifting, and how sometimes it can be more trouble than it's worth. Eventually Neil decided that he would charge friends full price and then give them a thank-you gift, like a box of chocolates. He decided that the manner in which he gave the gift was highly important and that he needed to be consistent, so he would prepare a standard letter explaining that the gift was a token of his appreciation for their business and that it was something he did for all his special friends. In effect, he put in place a gifting policy, knowing that when friends met at his Christmas party, no one would trade stories to find that someone else had received a bigger discount.

How to respond when you are "put on the spot"

In our speeded-up world, I am often asked to give an instant quote, when no two jobs are ever the same. Despite pressure to give a hasty quote "off the top of my head," I do not give in, because invariably those are the jobs I'll end up doing for nothing!

If the client is in a hurry, the brief is often rushed and essential details overlooked—and it is these oversights that have the greatest potential to cause misunderstandings. The dilemma if you increase the quote is that you may feel you risk client disappointment and may appear unprofessional—but not recovering the extra expenses will erode your profits.

If you want to maintain your integrity and leave clients wanting to do business with you again, you will find there is no substitute for going through the quoting process thoroughly. Put everything in writing, and despite the pressure, be firm and ask for the time you require: "I need to check a few details first; let me get back to you on that. When do you need to know by?"

Most often, the need to know is not a matter of life or death, although your customer may be acting as if it is. Personally, I prefer not to work with panic merchants; I don't like their style—although in the right circumstances, a desperate client can be very profitable!

"Give them no time to think" is a common business tactic, one often used by bullies. First, ask yourself: why is it so urgent? How long did it sit on their desk before they phoned and asked for your help? What cost-saving measures did they take to let things get as bad they are now? How inadequate have they been at managing the project?

I've often appreciated this quote by Rollo May: "It is an old ironic habit of human beings to run faster when we've lost our way."

My reference point is always: "Do I want to do this business at all?" I'd sooner do no business than do work that costs me money.[1] You might feel differently, and you are free to ignore my advice, especially if you're in the business of "rescuing" people. It can feel good; I've known the feeling myself, but have finally learned better!

---

1. Those with a large investment in capital equipment may want the work in order to churn some cash flow (albeit negative). Others may wish to work (unprofitably) to maintain visibility in the marketplace until things improve. Some will average the effects of good and bad months over the year. This comes back to your overall financial situation and the decisions you made about work, money, and the role of business in your life, as discussed earlier.

How to manage a client who quibbles over the bill

It happens! Things don't always go according to plan, and then there are some people who will always find something to quibble about, no matter how clear you thought you made things.

Obviously, if there's a major problem, sort it out first. A satisfied customer should be your number-one priority—then fix the bookkeeping. In each situation, assess the customer and the validity of their complaint then decide what your long-term intention is. Do you want to keep this customer and continue doing business with them? If you do, write those costs off to experience.

Otherwise, hang tough. Ask for what you believe you are entitled to, and if they don't like it, be prepared to watch them go. You know the possible consequences; take a deep breath, then face them. Every business has had clients that have mucked them around and not been worth the bother (except to teach us that we don't want to be treated like that anymore).

Just remember the Pareto Principle: 20 percent of your customers will be responsible for 80 percent of your profits. Likewise, 20 percent of your customers will be the cause of 80 percent of your grief! Concentrate your efforts on the 20 percent offering the biggest return on effort. Eliminate the 20 percent that annoy you.

What to do when confronted with a "deal-breaker"

> *Making money doesn't oblige people to forfeit their honor or their conscience.*
>
> —Baron Guy de Rothschild, banking magnate

A deal-breaker is someone who tries to renege on an agreement after it has been made. The best prevention is to get everything down in writing, but sometimes unforeseen things do crop up and genuine misunderstandings do occur. Worse still, you find you are dealing with someone who is trying to deceive you, and the person you thought you knew turns into another being altogether—a shape-shifter is what I call this type.

Sadly, I've been in this situation a couple of times, and though under normal circumstances I'd do almost anything to avoid conflict, these times I've held fast and stood my ground.

A multinational research company contracted me to run a short course. Our arrangement was that I would attend a three-day training course and undertake all the learning in my own time, with no pay. But when they sold a course, I would receive a percentage for facilitating it. Basically, I took all the risk, but stood to gain if the course took off in large numbers. The deal worked amicably the first few times, with six to eight students per course, but a problem arose when the company sold a course involving twenty students—then they suggested it might be a good idea to lower my percentage. Just when it was my turn to get a payback on the chance I took with them, they wanted to reduce my commission. *No deal!* I stood my ground, the course went ahead, and I received what I was entitled to.

Another deal-breaker happened with a client who always worked to frantic deadlines. On this particular project, he told me to go ahead and arrange the printing for a job. A day later, the client phoned to make some changes, by which time it was too late: the brochures were already done. The client then denied having told me to get them printed. In the end, it proved to be a cheap lesson, and since then, I always insist that clients sign the printing-approval form first. There are no exceptions to the rule; there's no point in putting systems in place to stop these things from happening if you're not actually going to follow them.

My final piece of advice is:

- Put the terms of every deal in writing. If you have a minimum charge, make sure your customer knows about it. Write down the decisions along with your obligations, commitment, and guarantees, and get them to sign their approval.
- Be specific about what you are quoting on. If there are any changes, list what they may be and how they might affect your quote. Pay attention to the details.
- Put in place systems for the benefit of the client and yourself.
- If the system works, stick with it. If you find a loophole, change the system, so it doesn't happen a second time.
- Try not to deal with clients who lack integrity in the first place.

Chapter 9

Focus on Your Dream

It's a funny thing about life: if you refuse to accept anything but the best, you very often get it.

—W. Somerset Maugham

Have you ever worked out how much money you have "left on the table" just because you never asked for it?

You've probably seen the sort of advertisements that investment companies like to use; in graph or table form, they show how savings reinvested with compound interest mount up—how $100 a week earning compound interest of 8 percent becomes nearly $110,000 in just ten years.

Add up all the hours, weeks, months, and years of cheap deals you've been doing, and estimate the opportunity cost of all that unclaimed money. Now resolve not to let it continue.

Focus on how good the future will be—how much better you'll feel. To help keep you motivated, place a reminder of the real reason you go to work somewhere you won't forget—near your diary, computer, or phone. It could be a picture of your child with the crooked teeth that you want to have straightened, or your dream home, or the feeling of pride and accomplishment you'll get when you bank the money—whatever it takes for you to resist the temptation to belittle your worth the next time you quote on a job.

Know that you are worth it

Go on, do it...the worst might never happen."
—Doreen Francis, my mother

Once you've set your price, be strong. Remind yourself that you are worth it. If the customer wants you, they're not going to get you any other way.

I was wearing a beautiful pink stone bracelet, when it got caught in some clothing and the setting was damaged. I knew the stones weren't valuable, so I continued to wear the bracelet until eventually one stone fell out. Do you think I would have done that if it had been a valuable pink diamond? Of course not. I would have gotten it fixed right away.

Make yourself valuable. Put a high price on your worth, and you may find that you get treated better.

I often quote the story about Picasso, who was asked to paint a picture for a customer. He spent ten minutes at the easel, then asked for $1,000. The customer was shocked and said, "But it only took you ten minutes." Picasso replied, "Yes, ten minutes plus a lifetime of experience."

Become discerning, and be willing to give up clients you don't want, so you can find clients you *do* want. If you spend all your time with clients that are only willing to pay $10 an hour, when will you have time for those ready to spend $100?

Give yourself encouragement

I suck at putting on my shoes, and my socks don't work.
—Calvin, five years old

Learning to tie shoelaces annoyed the hell out of my five-year-old son, but eventually he mastered the art. Be patient. It may not be easy the first time you try, but keep at it. Keep reminding yourself of the rewards in store.

Promise yourself a reward the first time (or every tenth time) you achieve what you set out to do. If you knew you were going to get a nice pay-off for going outside your comfort zone, you might feel a whole lot more motivated to try a new approach.

If someone else can do it better than you, let them!

Most celebrities and sports stars have agents. The beauty of an agent is that they have specialist industry knowledge, negotiating expertise, and a network they can tap into. In theory, they get paid to do a better job than you could do for yourself. If you can find someone to act in that capacity and it is a win-win situation, where you will both be better off for the arrangement, then do it. Just be absolutely certain you get this deal right.

Though most of us aren't celebrities or sports stars, we may still be able to get someone else to act as our agent when we apply for work as a temp, a contractor, or an employee. One of the advantages of being represented by a third party is that they may be more objective in seeing your worth; they may also sing your praises and virtues in a way that you can't because modesty forbids.

Alternatively, now that you're conversant with the ins and outs of charging what you are worth, you may be happy to take a "regular job" and receive a wage or retainer, so that you only have to negotiate the terms of your business the day you accept the position, and at review periods thereafter.

Having read all that is involved in price-setting, you may even feel a greater appreciation for your boss's entitlement to earn a margin (a return on investment) on the wages she is paying you, because you know the responsibility for pricing her business at a profit rests ultimately with her.

Wealth is a journey, not a destination

You've got to be happy living this life; it's your life, so do not be limited by the opinions of others. You don't have to—and nor can you ever—please everybody.

Inevitably, when you sell your house, car, or boat, there will be someone who says, "You could have gotten much more than that." Put those comments out of your mind. It's easy for them to say, but if you were pleased to take the money, be happy!

Life doesn't have to be about more and more, just for the sake of it.

Irrespective of how many dollars you earn, your true worth far exceeds great wealth; but if not charging what you are worth is a self-esteem issue, you will feel so much better when you find yourself being paid what you are worth. Do it once or do it a thousand times, and then when you know you can, do what you like. There is no law that says you have to; it's only if you want to.

The journey to a million starts with one

I will end with this story: Before I started training for the New York marathon, I had a philosophical discussion with my eldest daughter, wondering which bit of the training would be the most challenging—the beginning, the middle, or the end. I found out the beginning was hard; it was the height of summer and I was hopelessly out of shape, still breastfeeding my four-month-old baby. Then I found the middle months in winter were hard, especially the 6:00 AM starts for the two-to-three-hour runs every Saturday. And then, guess what? The end was hard too—*especially* as my friend and I staggered into Central Park just at dusk with the ambulance sirens getting increasingly more frequent as people in an even worse state than us were being hauled off the course for medical attention. Somehow or other, we made it over the finish line, but none of it would have happened if we hadn't taken that first step...and then another one.

As I write this, I realize that it was interesting that I never considered the consequences of quitting. It's obvious to me now, had I given up on my goal, that *that* would have been the hardest bit of all. You owe it to yourself to experiment a little. Don't wait for the time to be right. (It never has been until now.) Go out today, do a wonderful job, and charge that person exactly how much you know you are worth.

Recommended Reading

Carlson, Richard. *Don't Worry, Make Money: Spiritual and Practical Ways to Create Abundance and More Fun in Your Life.* Hodder & Stoughton, 1997.

Chan Kim, W., and Renee Mauborgne. *Blue Ocean Strategy: How to Create Uncontested Market Space and Make Competition Irrelevant.* Harvard Business School Press, 2005.

Daly, John L. *Pricing for Profitability: Activity-Based Pricing for Competitive Advantage.* Wiley, 2001.

Dolan, Robert J., and Hermann Simon. *Power Pricing.* Free Press, 1997.

Koch, Richard. *The 80/20 Principle: The Secret of Achieving More with Less.* Nicholas Brealey Publishing, 2003.

McGraw, Phil. *Life Strategies: Doing What Works, Doing What Matters.* Hyperion, 2000.

Mohammed, Rafi. *The Art of Pricing: How to Find the Hidden Profits to Grow Your Business.* Crown Business, 2005.

Nagle, Thomas T., and Reed K. Holden. *The Strategy and Tactics of Pricing.* Prentice Hall; third edition, 2002.

Nemeth, Maria. *The Energy of Money.* The Ballantine Publishing Group, 2000.

Ries, Al, and Jack Trout. *Positioning: The Battle for Your Mind.* McGraw-Hill, 2000.

Shefrin, Hersh. *Beyond Fear and Greed: Understanding Behavioral Finance and the Psychology of Investing.* Oxford University Press, 2002.

Wilde, Stuart. *The Trick to Money Is Having Some.* White Dove International, Inc., 1989.

Glossary of Terms

Advocates: customers who speak so highly about your business that they bring in more new customers for you.

Benchmark: a standard or reference point by which you measure something. Benchmarks are set to gain perspective: to measure where you are and how you are doing against how you once were and how you want to be.

Break-even: the point at which costs are covered, but no profit is made.

Commodity: a product that is undifferentiated from those of competitors; for example, petrol, table salt, unbranded flour and sugar, electricity.

Commission: a percentage on sales; brokerage.

Contingencies: planned or unexpected things that might occur in the future. Used in the context of setting aside some money to allow for contingencies.

Conversion rate: refers to the number of sales leads or customer enquiries that are converted into actual sales. If ten people reply to an advertisement but only one person purchases something from you, then you have a conversion rate of one in ten.

Cost-plus pricing: the artless form of pricing where costs are calculated and a profit margin based on a percentage of the costs is added on top. An overly simplistic form of pricing that ignores the profit opportunity created by pricing fees on the customer's perception of value.

Customer churn: when used by a marketer, this phrase refers to new customers that come and go. The objective of a profitable business is to acquire new customers and retain them.

Customer segmentation: the process of grouping individual customers into segments according to shared characteristics; for example, customers can be segmented into groups by age, geographic location, buying patterns, income data, educational background.

Discount: a reduction in the amount or cost.

Elastic demand: demand is considered to be elastic when a change in price causes a change in sales volume; for example, a half-price offer on facelifts may cause an increase in demand (elastic demand), whereas a fifty-percent fee reduction on emergency amputations is not likely to increase sales (inelastic demand).

Financial independence: the state of being able to live the life you want without having to go to work for money. Usually involves the notion of making your money work for you.

Fixed costs: costs that remain the same regardless of the quantity of products sold or amount of business done; for example, the company car, office furniture, monthly phone rental, professional indemnity insurance.

Forensic accountant: a specialist branch of accounting used to detect fraudulent activities.

Franchise: a business arrangement between a manufacturer or marketing organization and a separate operator that grants the operator license to sell a product, service, or system according to the "rules" of the manufacturer or marketing organization.

Gross profit: an amount of money made when a product or service is sold for more than it cost to make or deliver. Please note the difference between gross profit and net profit.

Inelastic demand: where a change in price doesn't affect demand for the product or service.

Infrastructure (as in "business infrastructure"): the components of the organization that are necessary to keep the business in business; for example, a plumber simply needs himself and a tool kit to fix a broken pipe, but his business infrastructure may involve a vehicle to get him to the job, an accountant to do his books, a secretary to answer his calls, etc.

Lifetime value of a customer: this marketing term refers to the total value of a customer to a business over a lifetime period. Lifetime value is a measure of the potential value a customer may contribute to a business; for example, the lifetime value of the man who comes into a children's store to buy a pram wheel to put on his golf cart is insignificant compared to the lifetime value of a pregnant woman who plans to have eight babies.

Loss leader: a product or service that is offered to the customer at an extremely attractive price, often below cost; like bait, a loss leader is offered in the hopes of catching "the big one" later.

Market share: a marketing term where the economic value of an industry or product class equals one hundred percent and a company calculates what "share" of the business they own. When businesses try to grow by going after market share, they try to win business away from their competitors. Another way to grow business is to create a new market; for example, the liquor industry did this when they introduced pre-mixed drinks.

Net profit: the profit your product (or business) has made after you have deducted expenses and paid yourself. Net profit is gross profit minus what you have paid yourself. If you intend to sell your business you need to focus on net profit, this is the amount of profit the *business* made.

Opportunity cost: the *total* calculated cost of missing out on business; for example, if I normally make $20,000 in May, but I decide to take a long vacation this May, the opportunity cost of not being at work is $20,000 *plus* the cost of getting back the business from customers I may have lost because I wasn't there as usual.

Overhead: the general, continuing cost of doing business, such as electricity, rent, etc.; the operational costs of being in business.

Penetration pricing: a pricing strategy, usually offering a very low price, designed to win business from your competitors and gain market share.

Perceived value: the value of your product or service, as your customer sees it.

Prestige pricing: pricing in a way that appeals to those with discerning taste and lots of discretionary disposable income; where the customer pays for the privilege of exclusivity or enhanced quality.

Price ceiling: the highest price customers will willingly pay.

Price floor: the lowest price customers will willingly pay. Wine at $2 a bottle is no longer perceived as wine, but vinegar instead.

Price-skimming: the practice of charging a high price to begin with; often a short-term profit-taking tactic while there are customers willing to pay a premium price. It is only effective until more competitors appear in the marketplace and/or customers balk at the price.

Price policy: a set of conditions by which a company makes pricing decisions.

Price points: this refers to the psychology of the customer with regard to the price they are willing to pay for a product or service. The price point is the point at which the price has tipped the balance and gone too far, and the customer will no longer buy; for example, some people may go looking for a car under $10,000. Car marketers often arrange their advertisements according to different price points. Notice also the number of cars for $9,995, not $10,000.

Price war: a battle between competitive companies or products fought on price.

Pro bono: services provided for free or donated without charge. Pro bono legal services are provided free to a client with the proviso that only if the client wins their case, they will pay the agreed fees.

Pro rata: services charged by an agreed proportion relative to success. A pro rata coaching agreement might mean the customer agrees to pay the coach proportionately according to successful results. The coach thereby shares in both the risk and rewards of success or failure, as agreed.

Product differentiation: the practice of customizing the characteristics of a product or service in order to make it meet the needs of different type of customers.

Profit: the surplus amount of money left over when costs are deducted from the sales price of a product or service. If there isn't a surplus, the amount is called a loss. Be sure to note the difference between gross profit and net profit.

Retainer: an ongoing fee that is paid to retain your services.

Return on investment: a calculation that measures the financial return on money spent or invested; utilizes the principle that money should be made to work for you. For example, if I invest $1,000 in the bank and get $100 paid to me in interest at the end of the year, then my return on investment is 10 percent. Equally, if I put $1,000 into a marketing campaign and sell $10,000 of services, of which $3,000 was net profit, then the return on my investment is 300 percent.

Revenue: the amount of money coming in. Revenue is a collective figure that includes costs and profits without differentiation.

Royalties: a licensing fee (often a percentage) paid upon the sale of individual items or deals.

Subcontract: where a business gains business only to get another person or business to supply the required items or service. A business may subcontract work when it has more work than it can handle or when it needs to buy additional expertise.

Value pricing: the practice of setting prices based on value. For example, a computer programmer may find it better to be paid for the value of the problem he helped solve, rather than the amount of time he spent fixing the problem; or a sales trainer may set her fees on the anticipated value of the increased sales, rather than the time it took to run the training course.

Variable costs: the costs involved in making a product or delivering a service that vary according to the quantity manufactured.

978-0-595-38601-7
0-595-38601-6

Printed in the United Kingdom
by Lightning Source UK Ltd.
126408UK00001B/396/A